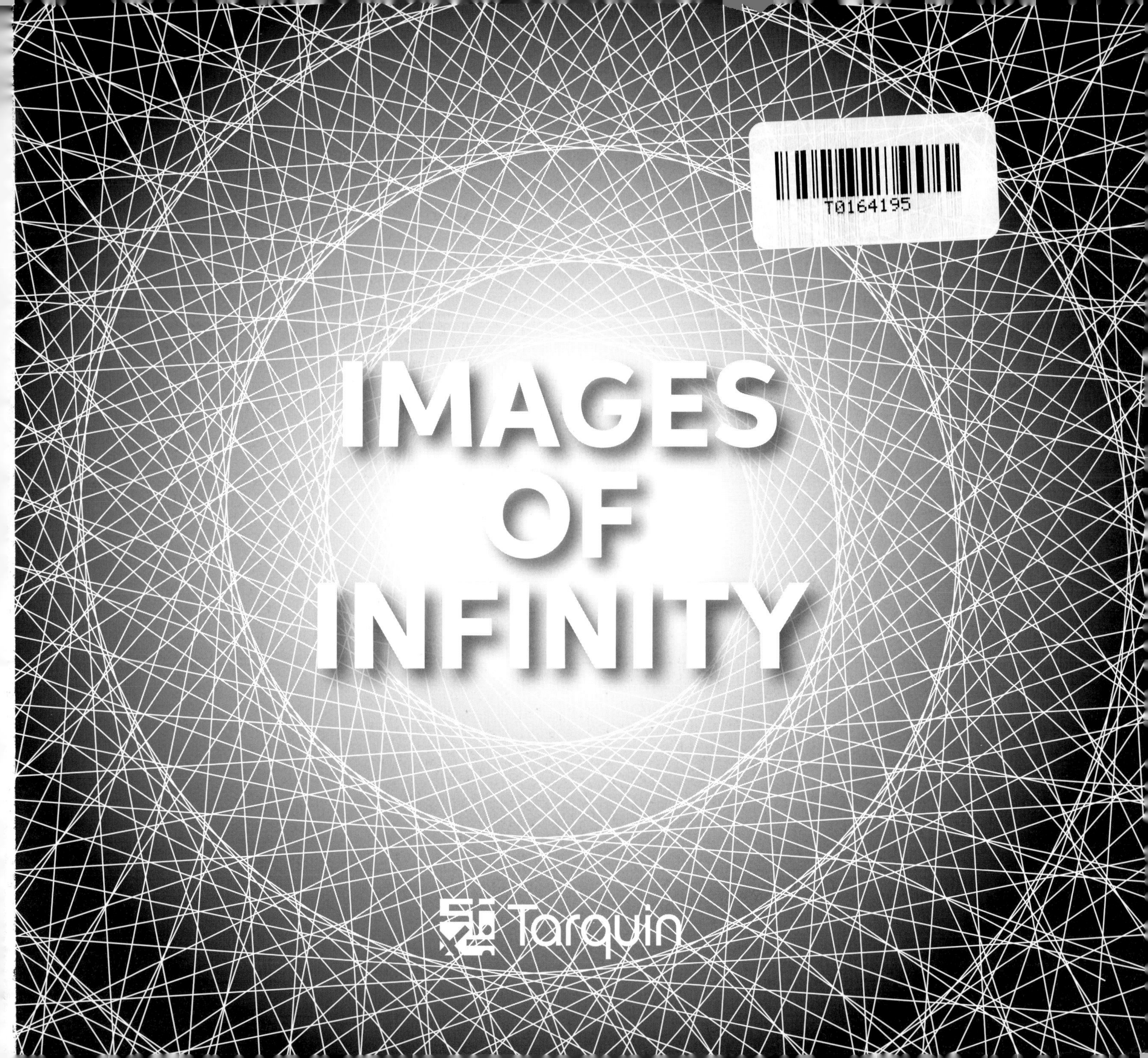
IMAGES
OF
INFINITY
Tarquin
T0164195

The original publication: © Leapfrogs Group 1984
Compiled and written: Ray Hemmings, Dick Tahta

Calligraphy: Gaynor Goffe
Illustrations: Klaas Bil (pp24, 39, 46, 62–65, 80–84, 87);
Vicky Squires (ppll, 86,88-90)

Original ISBN 0 905531 32 9
1995 edition ISBN 0 906212 89 8

**Acknowledgements**
"The Book of Sand" by Jorge Luis Borges is reproduced with permission of Penguin Books Ltd from The Book of the Sand, trans.

Norman Thomas di Ciovanni (Penguin Books 1979 pp87–91).



The woodcut Circle Limit 1 by M.C.Escher (1958) is reproduced (pl4) with permission. © SPADEM 1983.

Images from Italy reproduced on p51 and Danilo's painting (p50) are taken with permission from Sources magiques et poetiques des mathematicfues (Mouvement Cooperatif Educatif).

The photograph on p45 is reproduced courtesy of Barnaby's Picture Library.

**This edition**
© 2017 Leapfrogs Group
ISBN 978-1-911093-46-6 (paper)
ISBN 978-1-911093-47-3 (ebook)
Design: Trevor Bounford, after Paul Chilvers
Printed in the UK



Tarquin
Suite 74, 17 Holywell Hill
St Albans AL1 1DT
UK

www.tarquingroup.com

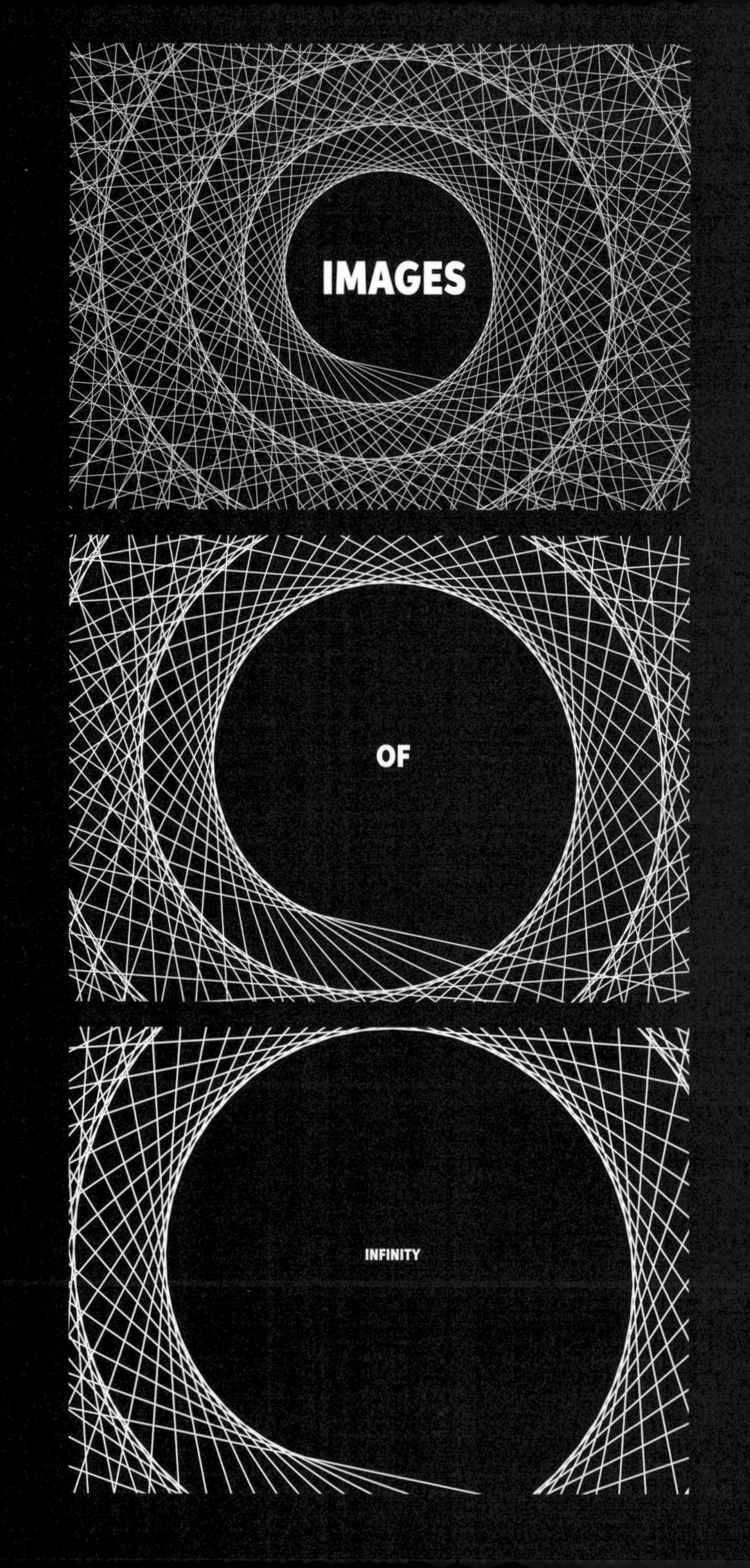

## Foreword

Who has not wondered if there is something beyond the universe? Or if there is a biggest number? Or if there is something smaller than the smallest thing? These are questions about different infinities. Exploring the paradoxes and riddles of what infinity means is the invitation contained in this book. And, more than that, in these pages is an invitation to engage in mathematics as a creative, collaborative endeavour, linked to our intuitions and emotions and of relevance to our lives.

Images of Infinity comes in sections that are loosely demarcated. Although there is a conceptual build, each section can also be worked on independently. There is content on each page that will repay the reader for careful and sustained engagement. There are insights and the potential for new awareness available throughout if, as a reader, you can give of your time and attention.

The ideas in this book are particularly suitable for joint activity, with friends or family. There may be some work needed to make sense of an unusual notation. There may be work needed to relate the text to the associated images. There may be a need for pen and paper and some extra thinking, or calculations, or diagrams, in order to convince yourself of an argument made in the text. This is a book that invites exploration.

The main text of the book begins (on p.8) with the notion of 'recursion', a theme through the early pages. The experiment with a mobile phone, on page 8, is well worth doing as a first activity and can be approximated by simply pointing a mobile phone, with a front facing camera, at a mirror. Getting comfortable with a range of images of recursion is an important step towards becoming acquainted with infinity. Another recurring theme is the surprising difference between the sums of these two sequences of fractions:

1/2+1/4+1/8+1/16+...      1/2+1/3+1/4+1/5+...

The difference is explored, among other places, in the Gog and Magog story on p.62, which makes accessible a rigorous argument as to the two totals.

One idea in the book that may be unfamiliar to some readers is that of 'number bases'. At the time of the first edition, number bases were in the UK curriculum and so may have been more familiar to the general public. On p.68 is an argument using base 9 compared to base 10 representations of numbers, that may help readers understand what bases mean. An example that may be recognised is that of binary or base 2; in base 2, the counting sequence begins: 1, 10, 11, 100, 101, 110, 111, 1000, 1001, etc. So, our (base 10) number 7, say, is represented as 111 in base 2. What we might think of as the 'tens column' represents the number of 2s, (the base). What we might think of as the 'hundreds column' represents the number of (22)s (the base squared). So, without counting up, we can 'read' 111 (base 2) as 1 x 22 + 1 x 21 + 1 = 4 + 2 + 1 = 7 (base 10).

Perhaps even less familiar, at the present time, is the use of different number bases to represent decimals. This is an idea that recurs in the later pages of the book. Bicimals (decimals in base 2) are introduced on page 41. In the same way that our 'hundreds' column can be viewed as the 'base 10 squared' column, so the 'tenths' column is 'one over the base' and the 'hundredths' column is 'one over the base squared'. In base 2 (bicimals) this means a number such as 0.11, is read as: 1 x (1/2) + 1 x (1/22) = ½ + ¼ = ¾ = 0.75 (base 10). On p.76, there is use of 'tricimal' notation (decimals in base 3), where 0.1 is 1/3 (base 10), 0.2 is 2/3 (base 10), 0.01 is 1/9 (base 10) and so on. There are patterns that become apparent through the use of these different bases that are much less obvious in our standard base 10. An interesting exercise is to translate or extend the arguments into base 10.

The activities and ideas in this book were put together by two of the most innovative mathematics educators of the last century. Each page has been designed with care and attention to detail and points to fundamental awareness(es) about the paradoxes and pleasures of dealing with the infinite. To engage actively in this book, therefore, is to benefit from the experiences, wisdom and insight of two great teachers of mathematics.

Alf Coles

These difficulties are real ... But let us remember that we are dealing with infinities and indivisibles, both of which transcend our finite understanding .. In spite of this men cannot refrain from discussing them.

GALILEO

FROM THIS
MONUMENT
HANGS
ETERNITY

PART I
PART I
PART I
PART I
PART I

TIONS
TIONS
TIONS
TIONS

## Recursion

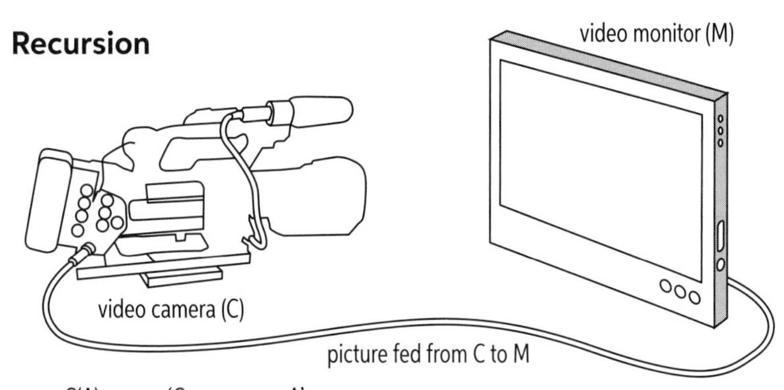


C(A) means 'Camera sees A'

and M(A) means 'Monitor shows picture of A'.

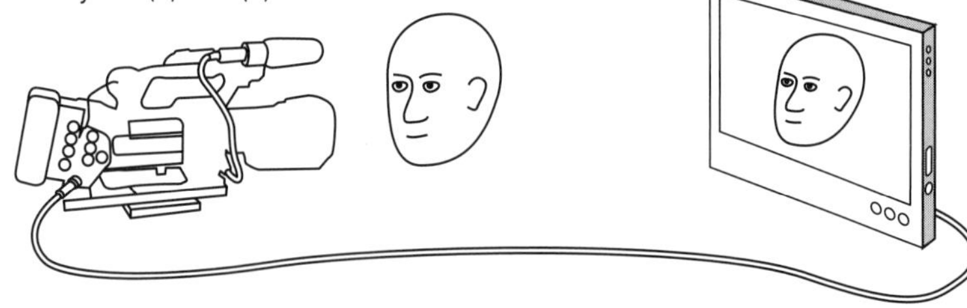

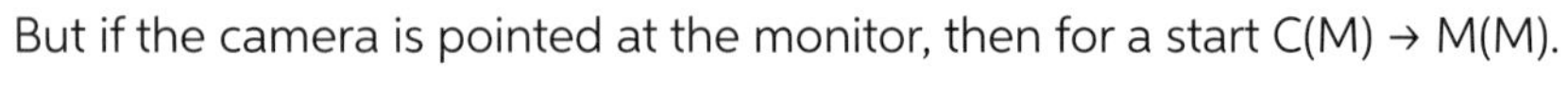


But now the camera sees, not just M, but 'M showing M'.

So $C(M(M)) \to M(M(M))$.

And now it sees 'M showing M showing M'

$$C(M(M(M))) \to M(M(M(M))).$$

And so it goes on, indefinitely—
but it all happens instantaneously!

In computing language, there is an 'infinite loop'

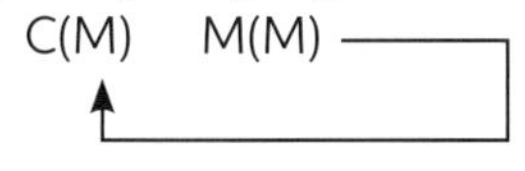


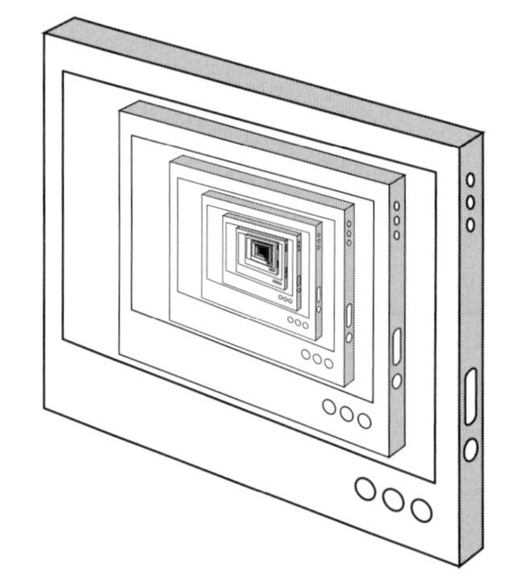

This endless feeding back is called 'recursion'.

I am a smartphone
taking a picture of myself –
taking a picture of myself –
taking a picture of myself –
taking a picture of myself –
taking a picture of myself –
taking a picture of myself –
taking a picture of myself –
taking a picture of myself –
taking a picture of myself –
taking a picture of myself –
taking a picture of myself –
taking a picture of myself –
taking a picture of myself –
taking a picture of myself –
taking a picture of myself –
taking a picture of myself –
taking a picture of myself –

## Arithmetic recursion

- We have a function machine. It multiplies any number fed into it by 0.5.

  Now we fit a loop to this machine so that the output is fed back into the machine again—and again, and again . . .

- The numbers which the machine now produces, 6, 3, 1.5 ... (a half of a half of a half of …) are something like the picture of the picture of the picture of ... on the TV monitor

- But the picture on the monitor contains the whole set of 'pictures of'. To make it more like this we could plug in a 'totalizer' which keeps a running total of the outputs.

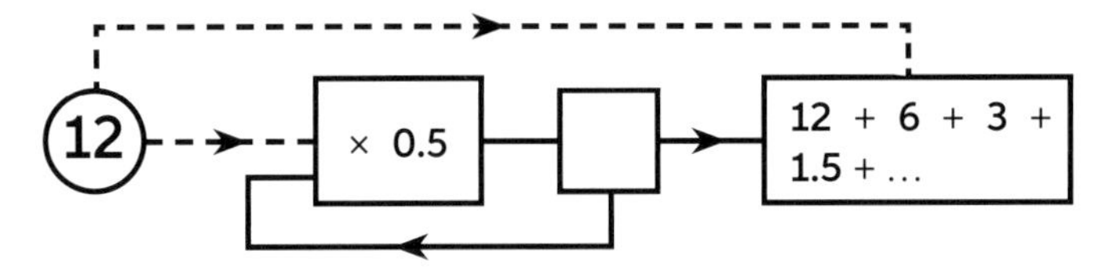


  Will this total get indefinitely large?

- Another way of getting the same effect is by this arrangement:

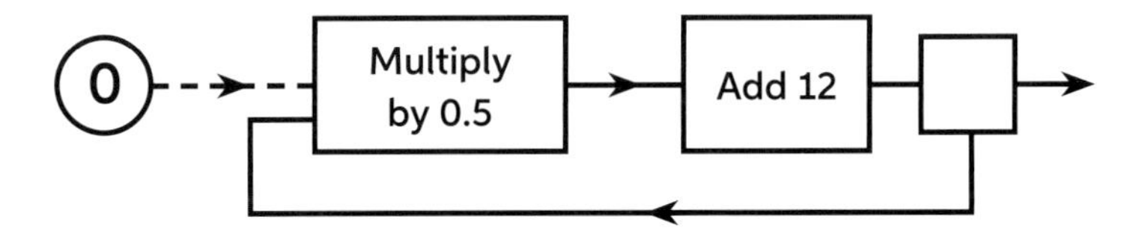


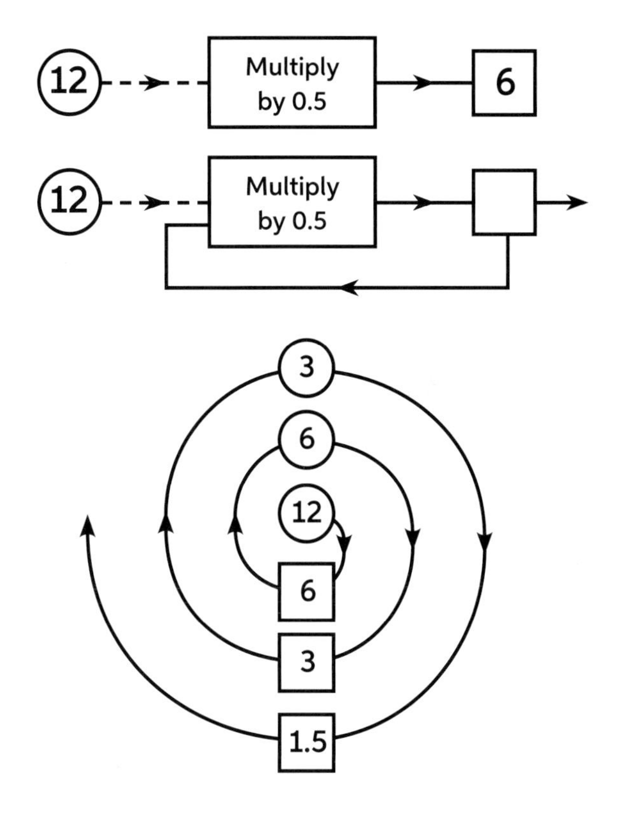

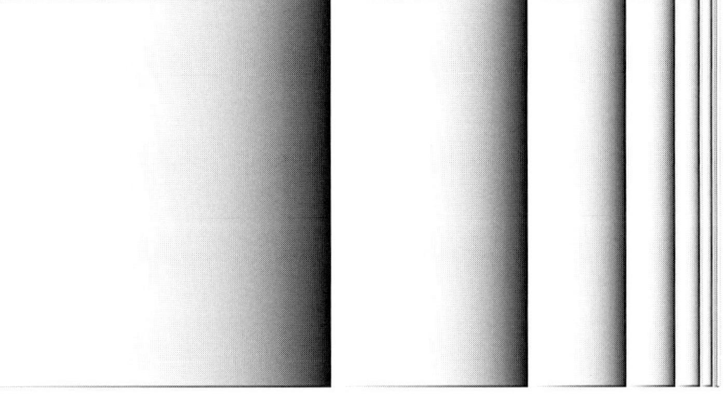

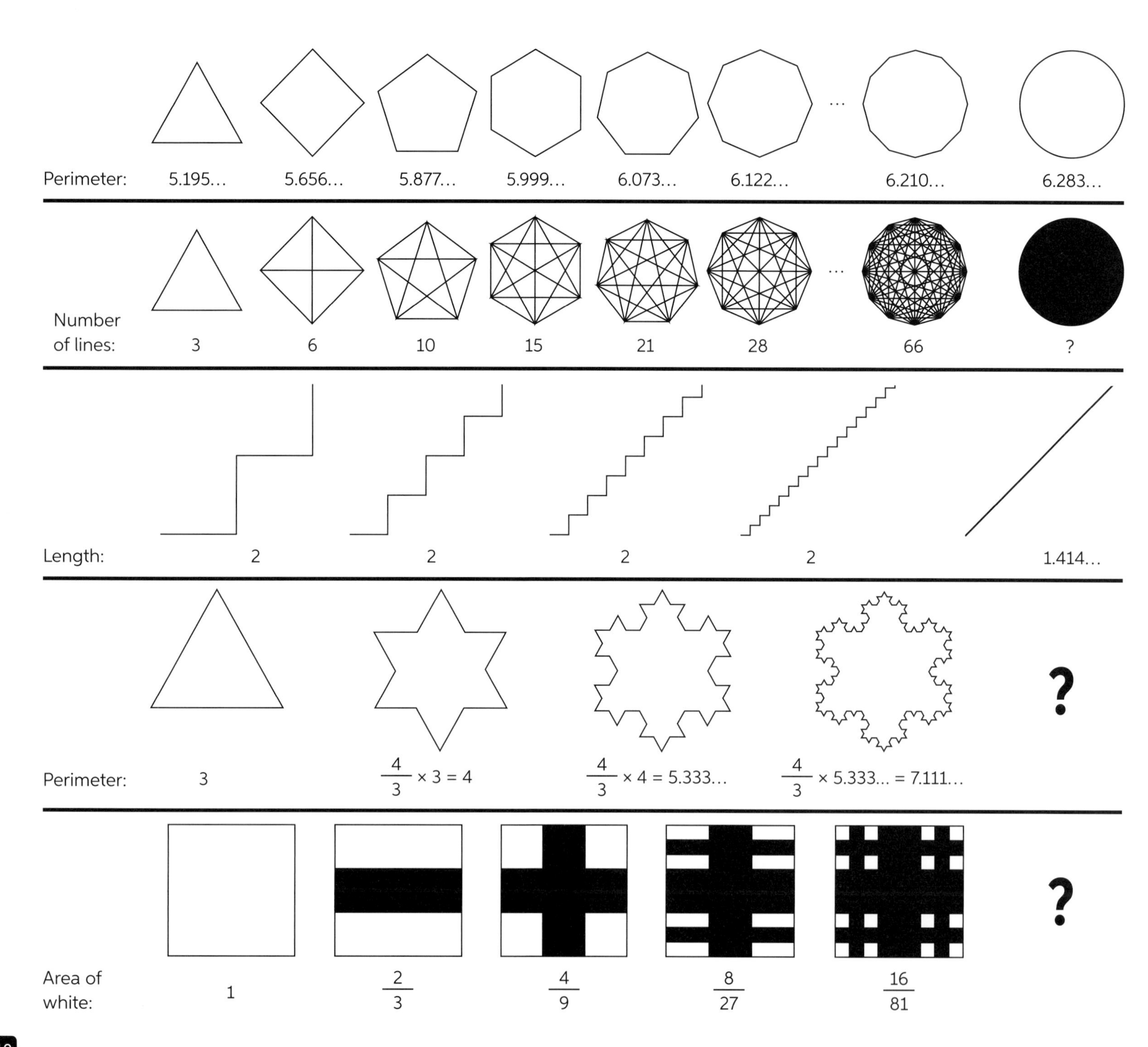
Perimeter:
5.195…
5.656…
5.877…
5.999…
6.073…
6.122…
6.210…
6.283…
Number of lines:
3
6
10
15
21
28
66
?
Length:
2
2
2
2
1.414…
Perimeter:
3
$\frac{4}{3} \times 3 = 4$
$\frac{4}{3} \times 4 = 5.333\ldots$
$\frac{4}{3} \times 5.333\ldots = 7.111\ldots$
?
Area of white:
1
$\frac{2}{3}$
$\frac{4}{9}$
$\frac{8}{27}$
$\frac{16}{81}$
?

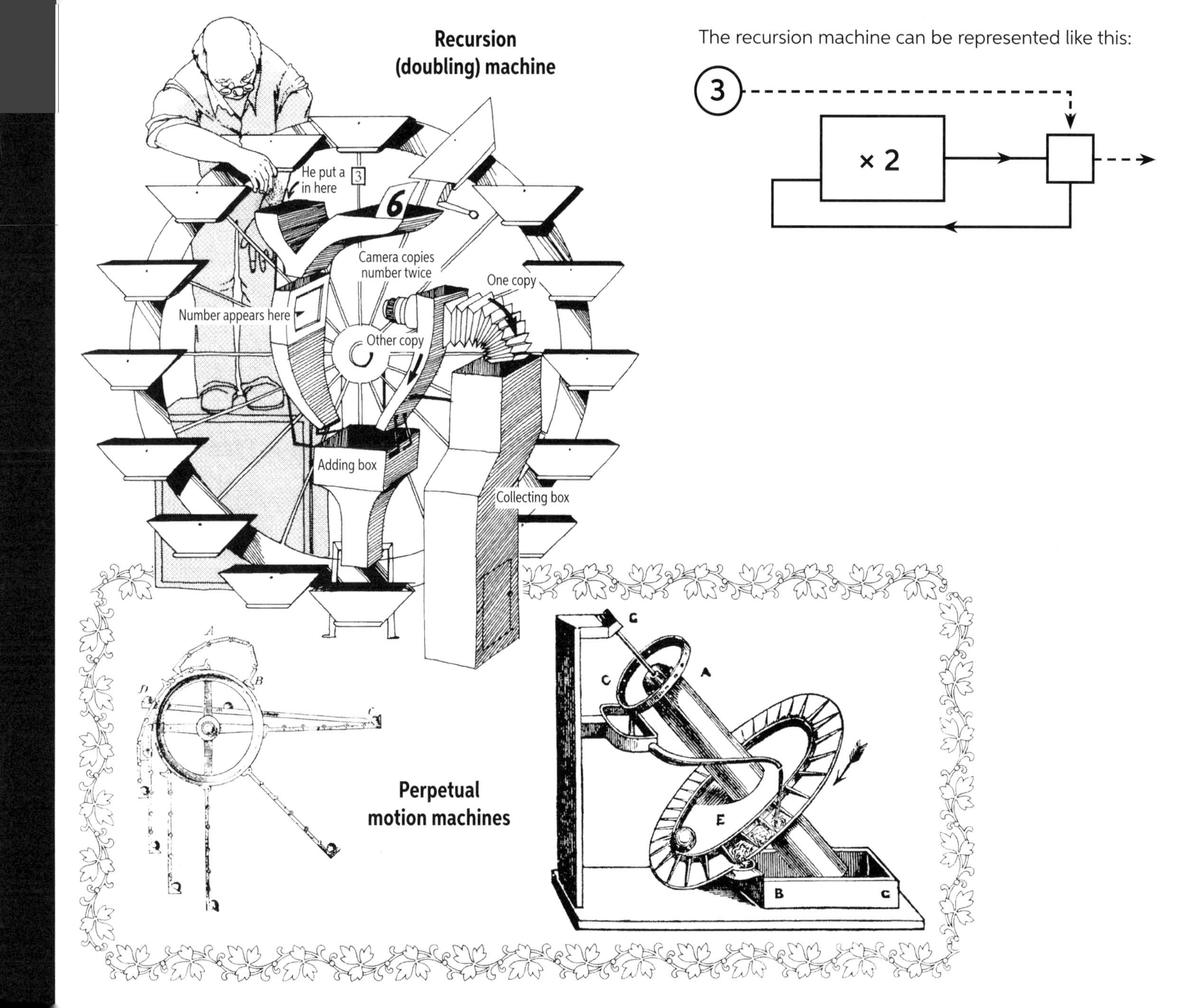

Recursion (doubling) machine
He put a 3 in here
6
Camera copies number twice
One copy
Number appears here
Other copy
Adding box
Collecting box
The recursion machine can be represented like this:
3
× 2
Perpetual motion machines

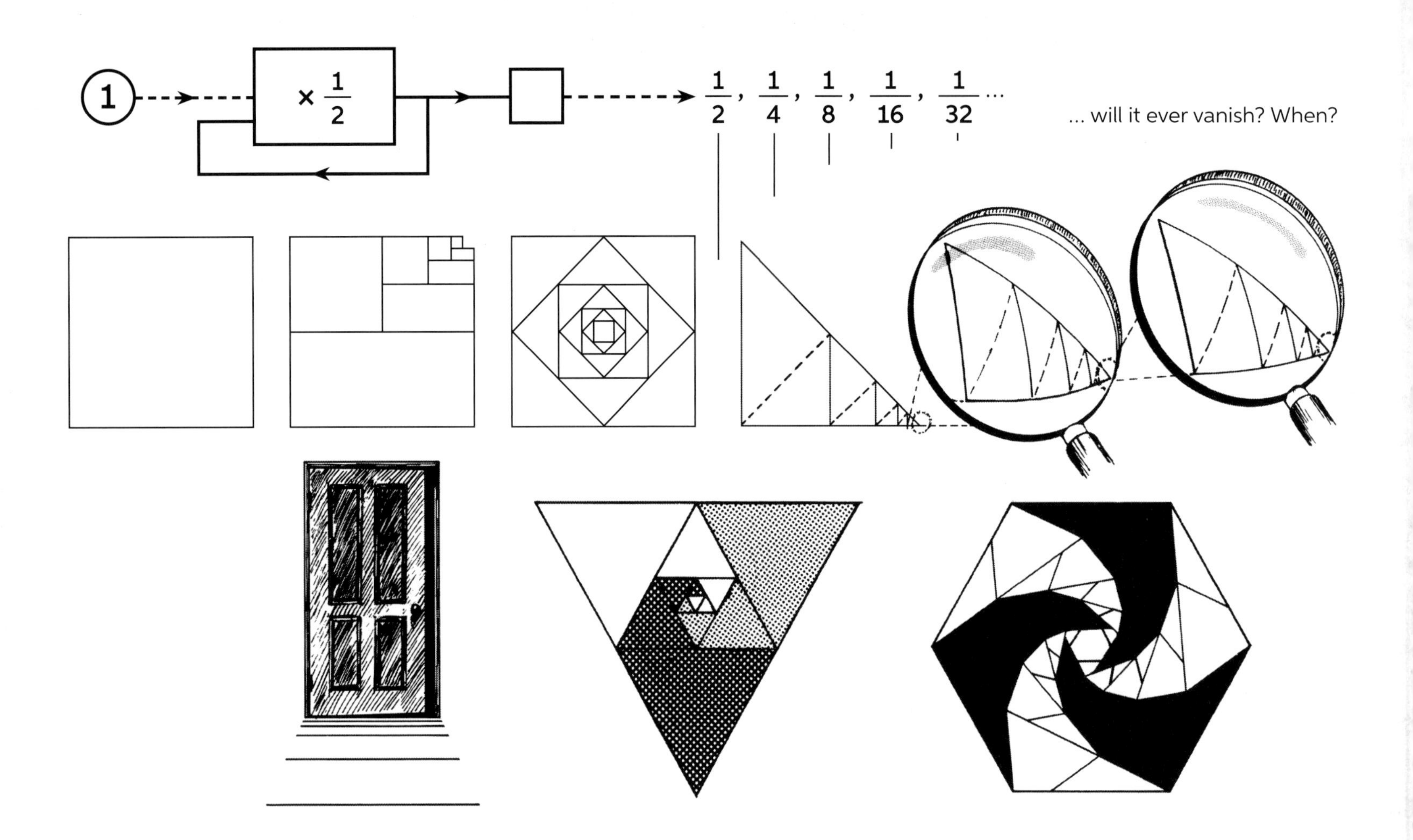


**You'll never get out of the room ...**
How far is the door? Eight feet? If you got half-way there, you'd still have half the distance (4ft) to go. And if you went half that distance you'd still have half that distance to go. Thinking of it this way, you can see that there is always a half-distance to go. So you'll never reach the door! In fact the same argument, applied to the half-distance (4ft), proves you'll never get that far ... nor will you get a quarter the way... and so on... you can never move at all!

The story on the left was first told by Zeno 25 centuries ago... Modern scientists tell a similar story about radioactive isotopes. For any particular element, half of the radioactive atoms decay into some other isotope in a certain period (called the half-life of that element). For example, half the carbon-14 atoms in a piece of dead material change into nitrogen-14 atoms after about 5720 years. But half of them are still there. In another 5720 years half of that half will have decayed. If it went on like this, would the carbon ever completely disappear?

*The infinite exists in the imagination:*
*not the object of knowing imagination*
*but of imagination that is uncertain about its object,*
*suspends further thinking*
*and calls infinite all that it abandons.*

*Just as sight recognises darkness*
*by the experience of not seeing*
*so imagination recognises the infinite*
*by not understanding it.*

[PROCLUS 412–85 AD]

*Cantor [1845–1918], having proved that*
*the infinity of points within a square is equal to*
*the infinity of points on one of its sides,*
*wrote to his friend Dedekind:*
*'I see, but I do not believe it.'*

*By adding continuously to a finite size*
*I will pass any limited size.*
*By subtracting, I will in the same way*
*leave one which is smaller than any other.*

[ARISTOTLE 384–322 BC]

The *infinitive* does not denote any preciſe time, nor does it determine the number, or perſons, but expreſſes things in a looſe indefinite manner; as, *to tea:h*, &c.
In moſt languages, both antient and modern, the *infinitive* is diſtinguiſhed by a termination peculiar to it; as τυπτειν in the Greek, *ſcribere* in the Latin, *ecrire* in the French, *ſcrivere* in the Italian, *&c.* but the Engl:ſh is defective in this point; ſo that to denote the *infinitive*, we are obliged to have recourſe to the art cle *to*; excepting ſometimes when two or more *infinitives* follow each other.
The practice of uſing a number of *infinitives* ſucceſſively, is a great, but a common fault in language; as, *he offered to go to teach to write* Engliſh.—Indeed, where the *infinitives* have no dependence on each other, they may be uſed elegantly enough; as, *to mourn, to ſigh, to ſink, to ſwoon, to die.*
INFINITY, the quality which denominates a thing *infinite.* The idea ſignified by the name *infinity* is beſt examined, by conſidering to what things *infinity* is by the mind attributed, and how the idea itſelf is framed: finite and infinite are looked upon as the modes of quantity, and are attributed primarily to things that have parts, and are capable of increaſe or diminution, by the addition or ſubtraction of any the leaſt part. Such are the ideas of ſpace, duration, and number.—When we apply this idea to the ſupreme being, we do it primarily in reſpect of his duration and ubiquity; and more figuratively, when to his wiſdom, power, goodneſs, and other attributes, which are properly inexhauſtible and incomprehenſible: for when we call them *infinite*, we have no other idea of this *infinity*, but what carries with it ſome reflection on the number, or the extent, of the acts or objects of God's power and wiſdom, which can never be ſuppoſed ſo great, or ſo many, that theſe attibutes will not always ſurmount and exceed, though we multiply them in our thoughts with the *infinity* of endleſs number. We do not pretend to ſay, how theſe attributes are in God, who is infinitely beyond the reach of our narrow capacities; but this is our way of conceiving them, and theſe are our ideas of their *infinity.*
We come by the idea of *infinity* thus: Every one that has any idea of any ſtated length of ſpace, as a foot, yard, *&c.* finds that he can repeat that idea, and join it to another, to a third, and ſo on, without ever coming to an end of his additions. From this power of enlarging his idea of ſpace, he takes the idea of infinite ſpace, or immenſity. By the ſame power of repeating the idea of any length or duration we have in our minds, with all the endleſs addition of numb:r, we alſo come by the idea of eternity.
If our idea of *infinity* be got, by repeating without end our own ideas, it may be aſked, Why do we not attribute it to other ideas, as well as thoſe of ſpace and duration; ſince they may be as eaſily, and as often, repeated in our minds, as the other? yet nobody ever thinks of infinite ſweetneſs, or whitenels, though he can repeat the idea of ſweet or white, as frequently as thoſe of yard or day? To this it is anſwered, that thoſe ideas which have parts, and are capable of increaſe by the addition of any parts, afford us by their repetition an idea of *infinity*; becauſe with the endleſs repetition there is connected an enlargement, of which there is no end: but it is not ſo in other ideas; for, if to the perfecteſt idea I have of white, I add another of equal whiteneſs, it enlarges not my idea at all. Thoſe ideas, which conſiſt not of parts, cannot be augmented to what proportion men pleaſe, or be ſtretched beyond what they have received by their ſenſes; but ſpace, duration, and number, being capable of increaſe by repetition, leave in the mind an idea of an endleſs room for more; and ſo thoſe ideas alone lead the mind towards the thought of *infinity.*
We are carefully to diſtinguiſh between the idea of the *infinity* of ſpace, and the idea of a ſpace infinite.—The firſt is nothing but a ſuppoſed endleſs progreſſion of the mind over any repeated idea of ſpace: but to have actually in the mind the idea of a ſpace infinite, is to ſuppoſe the mind already paſſed over all thoſe repeated ideas of ſpace, which an end-

**Extracted from an 18th-century encyclopaedia**

XI-'58

Three points are chosen on the 'horizon'.

A line is drawn through each of these points.

After this, any crossing of two lines is joined to the third point on the 'horizon'.

The drawing is unfinished—there are still many two-line crossings to be joined to a third point.

Could the drawing ever be finished?

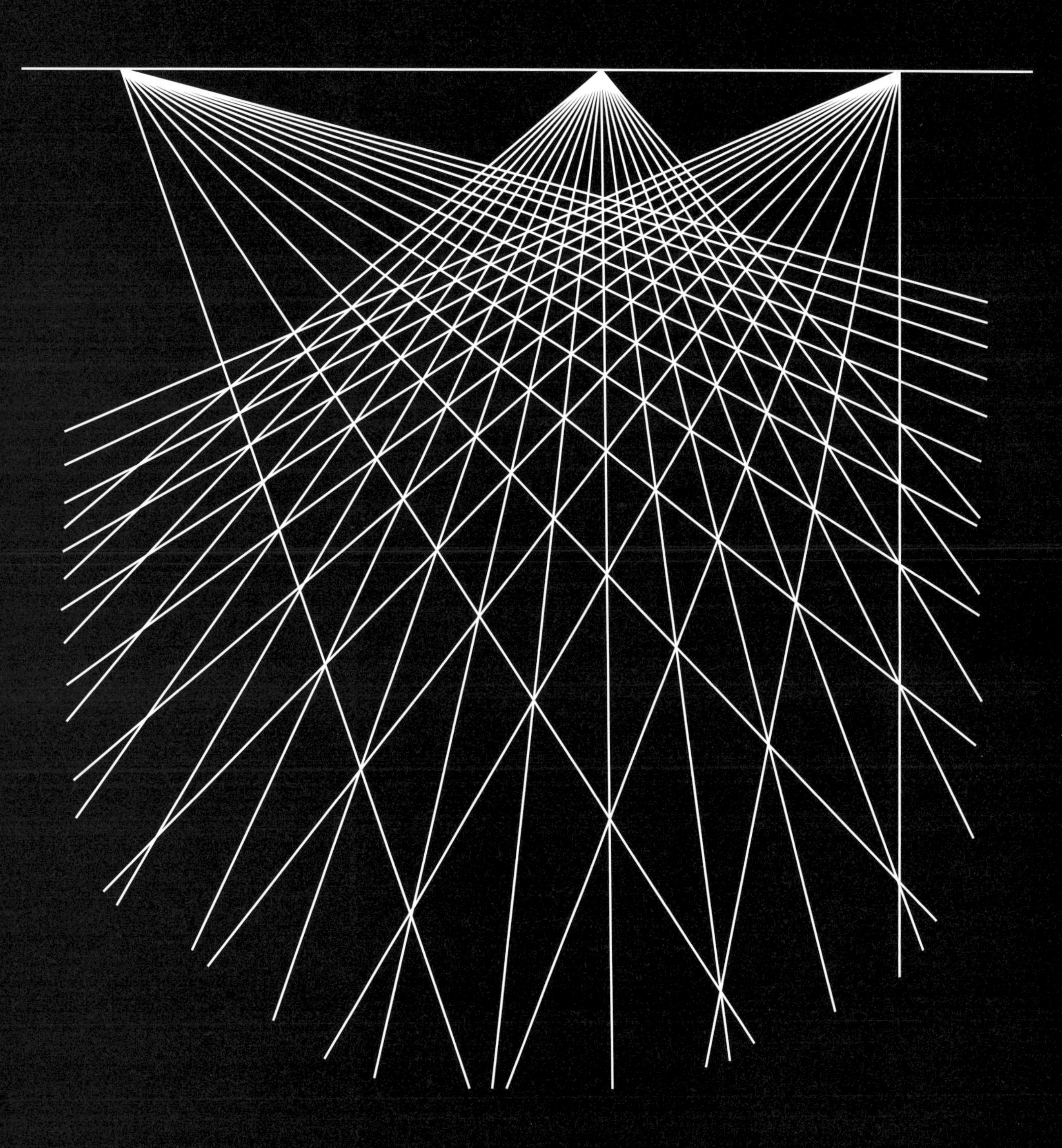

The three horizon points have been taken on a circle rather than a line.

The starting lines crossed inside the triangle. After that the crossings appear to stay inside.

Will the lines ever reach the sides of the triangle?

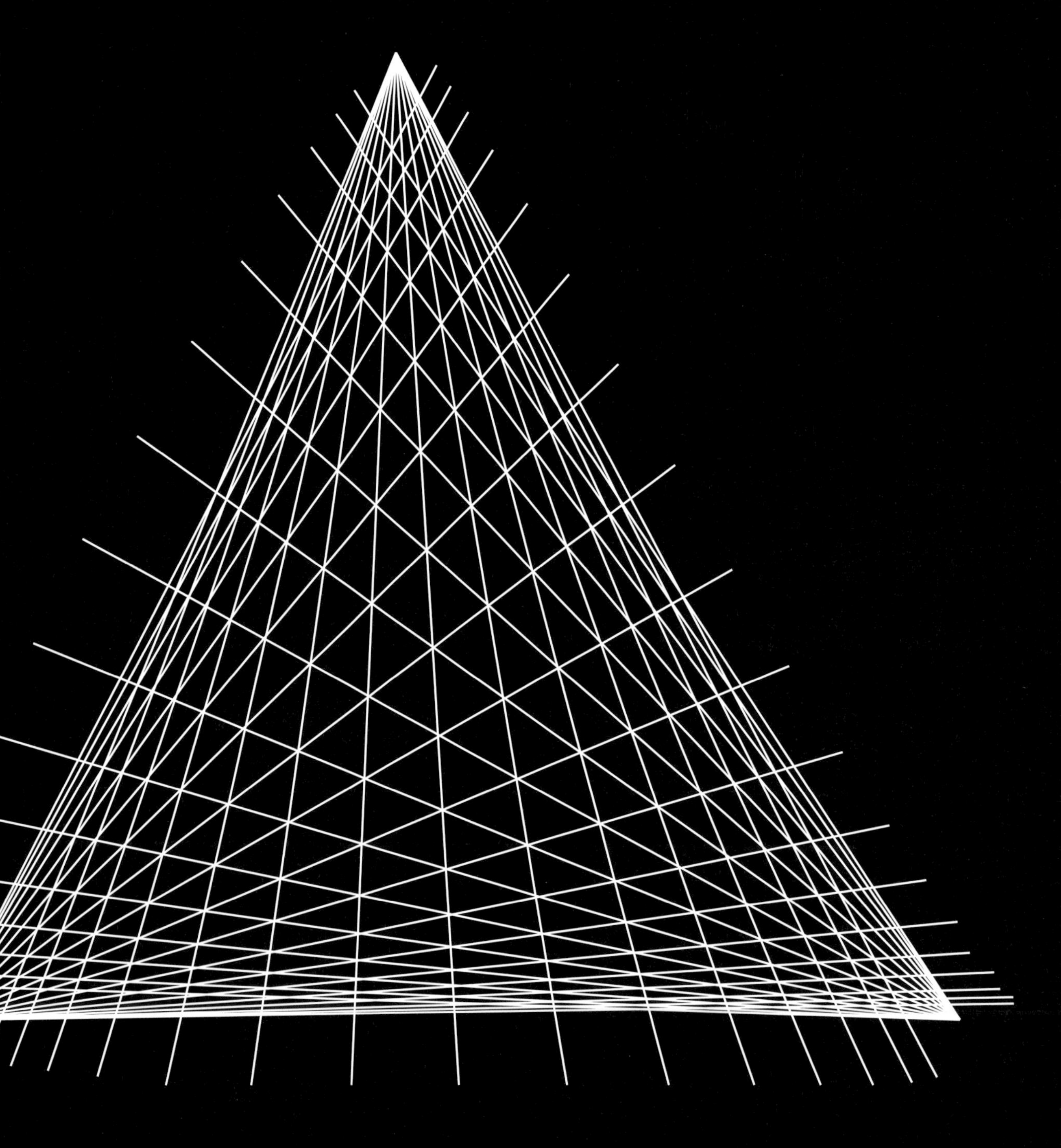

Most number series, which have a rule to get the next number, go on for ever ,

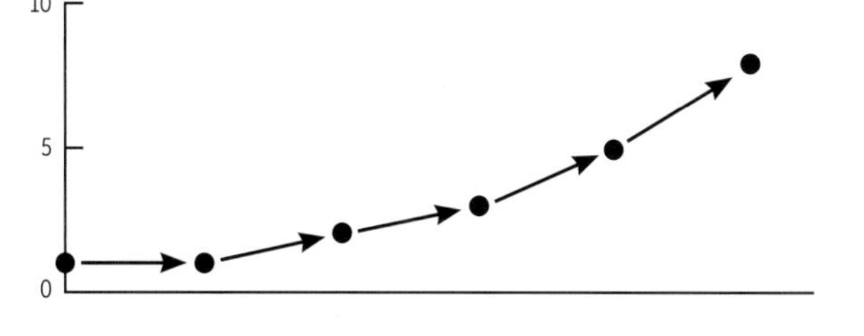


In some cases the numbers get larger and larger . . .

RULE: Multiply by 3 → 1, 3, 9, 27 …

RULE: Add 1 → 5, 6, 7, 8 …

RULE: Add last two terms together → 1, 1, 2, 3, 5, 8 …

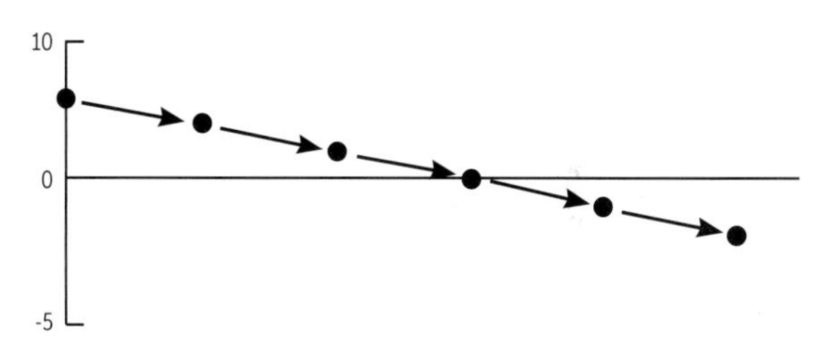


In other cases the numbers get smaller without limit

RULE: Subtract 1 → 3, 2, 1, 0, −1 …

RULE: Multiply by 3 → −1, −3, −9 …

RULE: Subtract the number of the term → 10, 8, 5, 1, −4 …

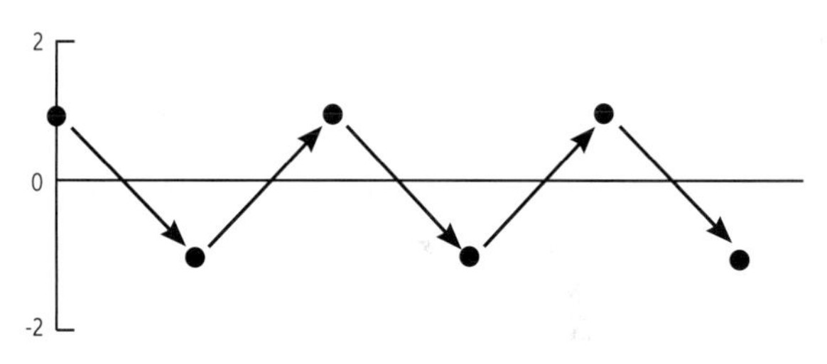


In yet other cases the numbers oscillate

RULE: Multiply by −10 → 1, −10, 100, −1000 … the swings get bigger and bigger

RULE: Multiply by −1 → 1, −1, 1, −1, ... the swings are fixed

RULE: Multiply last two terms together → 2, −3, −6, 18, −108, …

---

Some series increase or decrease continually—but not without limit

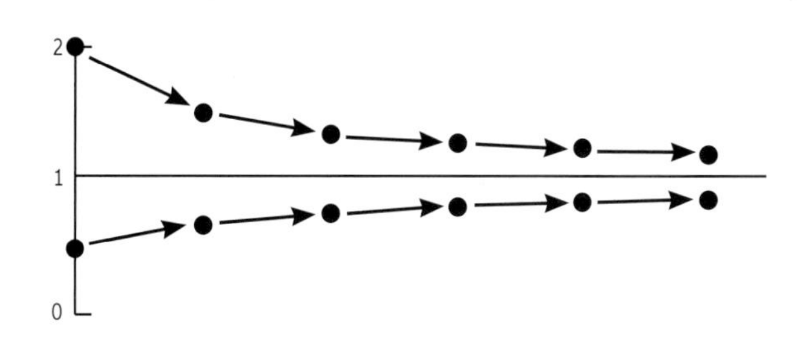


RULE:

Add 1 to numerator and add 1 to denominator → $\frac{2}{1}, \frac{3}{2}, \frac{4}{3}, \frac{5}{4} \ldots$

RULE:

The same (but a different starting point) → $\frac{1}{2}, \frac{2}{3}, \frac{3}{4}, \frac{4}{5} \ldots$

---

Imagine something shrinking, always shrinking—but never getting smaller than … so much. Can you?

Imagine something growing, always growing—but never getting larger than …

In the two series above the numbers get nearer and nearer to 1, but never actually get there! Will they ever get to within one millionth of 1?

Some examples: which seem to be 'levelling off

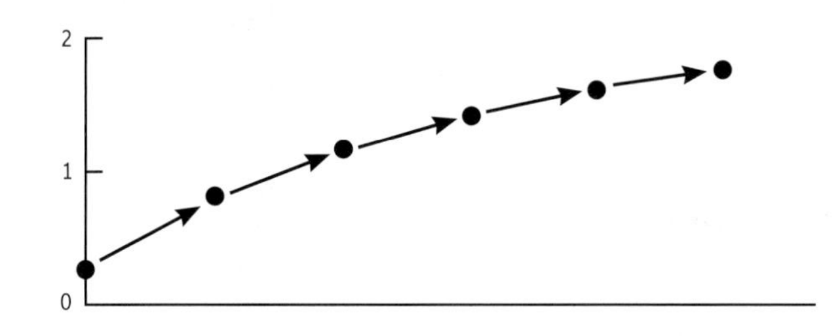


RULE: Add 3 to the numerator and add 1 to the denominator →

$$\frac{1}{4} \quad \frac{4}{5} \quad \frac{7}{6} \quad \frac{10}{7} \quad \frac{13}{8} \ \ldots$$

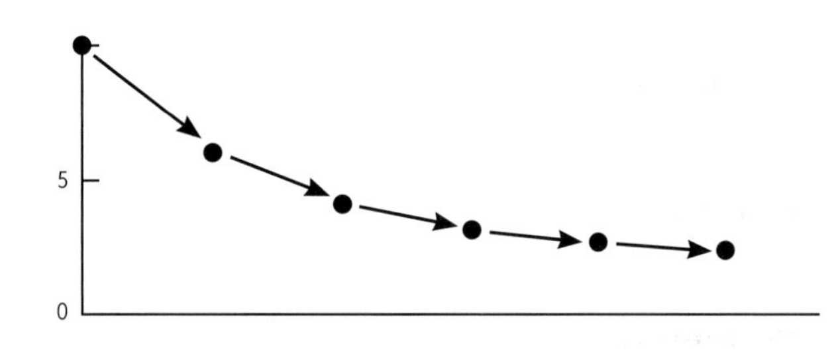


RULE: Take half, and add 1
10, 6, 4, 3, $2\frac{1}{2}$, $2\frac{1}{4}$ …

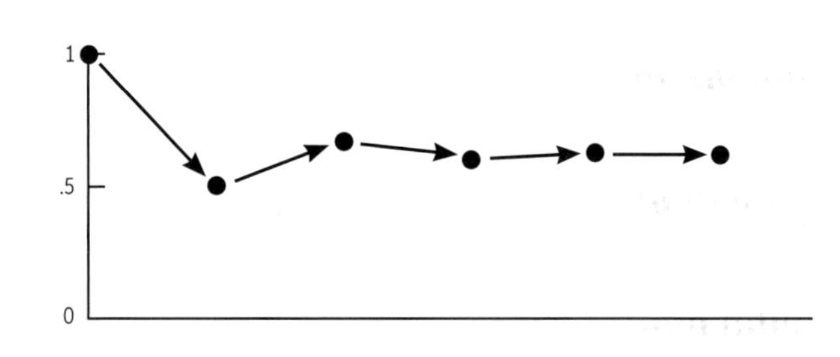


Oscillating to a limit:
RULE: Make fractions by successive terms of the Fibonacci series (that's when you add the last two terms together)

$$\frac{1}{1}, \frac{1}{2}, \frac{2}{3}, \frac{3}{4}, \frac{5}{5}, \frac{8}{13} \ \ldots$$

---

What happens if you write all the whole numbers starting at 2 in two rows. If a number is prime it goes in the top row. If not, in the bottom row. Pair the numbers off and write them as fractions:

2 3 5 7 11 13 17 19 23 …
4 6 8 9 10 12 14 15 16 18 20 21 22 …

$$\frac{2}{4} \quad \frac{3}{6} \quad \frac{5}{8} \quad \frac{7}{9} \quad \frac{11}{10} \quad \frac{13}{12} \quad \frac{17}{14} \quad \frac{19}{15} \quad \frac{23}{16} \ \ldots \ \frac{103}{40} \ \ldots \ \rightarrow$$

It helps to change the fractions into decimal form:
0.5, 0.5, 0.625, 0.7, 1.1, 1.08 …, 1.21 …, 1.26 …, 1.43 …, 2.575 …

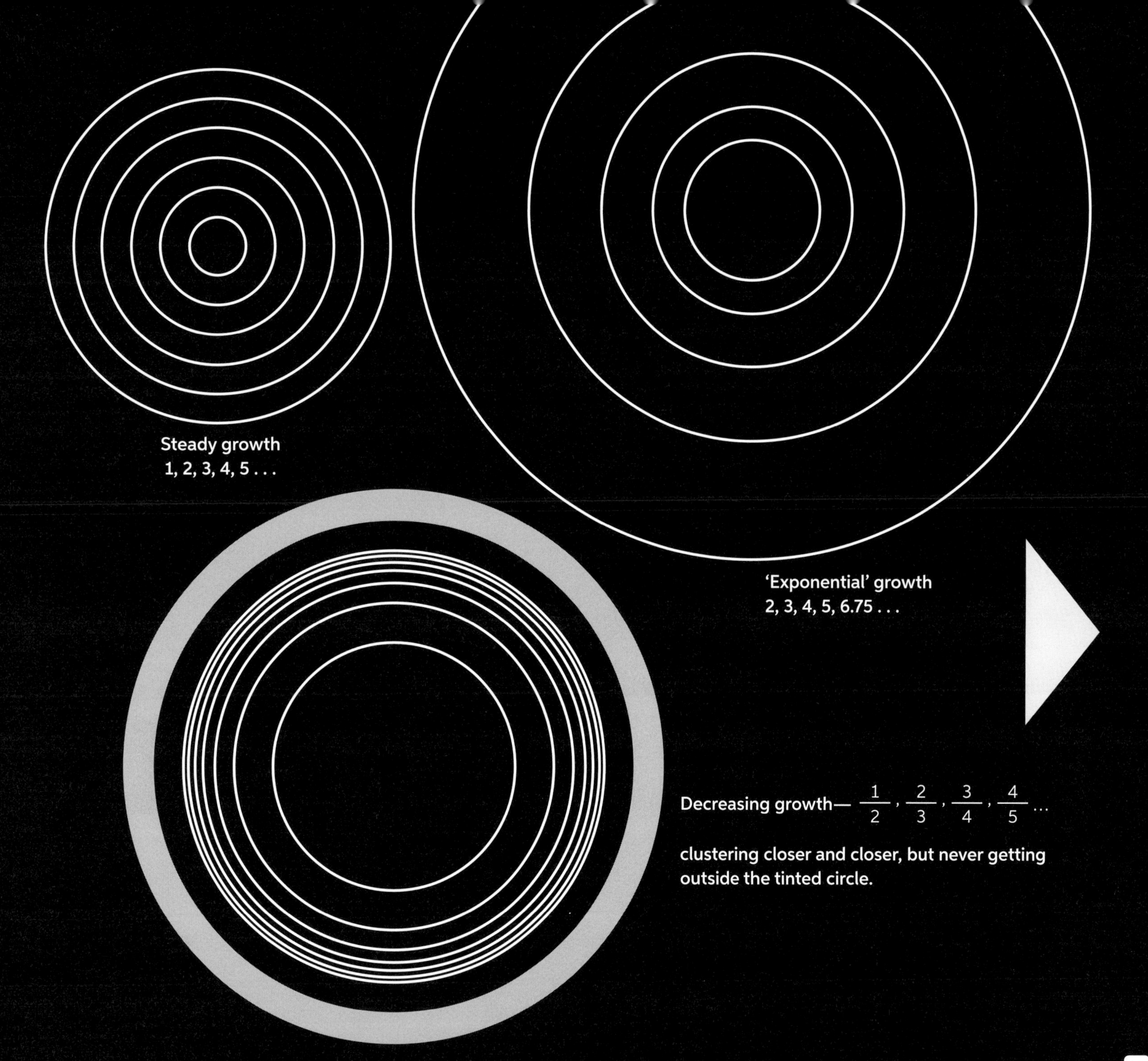
Steady growth
1, 2, 3, 4, 5 . . .
'Exponential' growth
2, 3, 4, 5, 6.75 . . .
Decreasing growth— $\frac{1}{2}, \frac{2}{3}, \frac{3}{4}, \frac{4}{5} \ldots$
clustering closer and closer, but never getting outside the tinted circle.

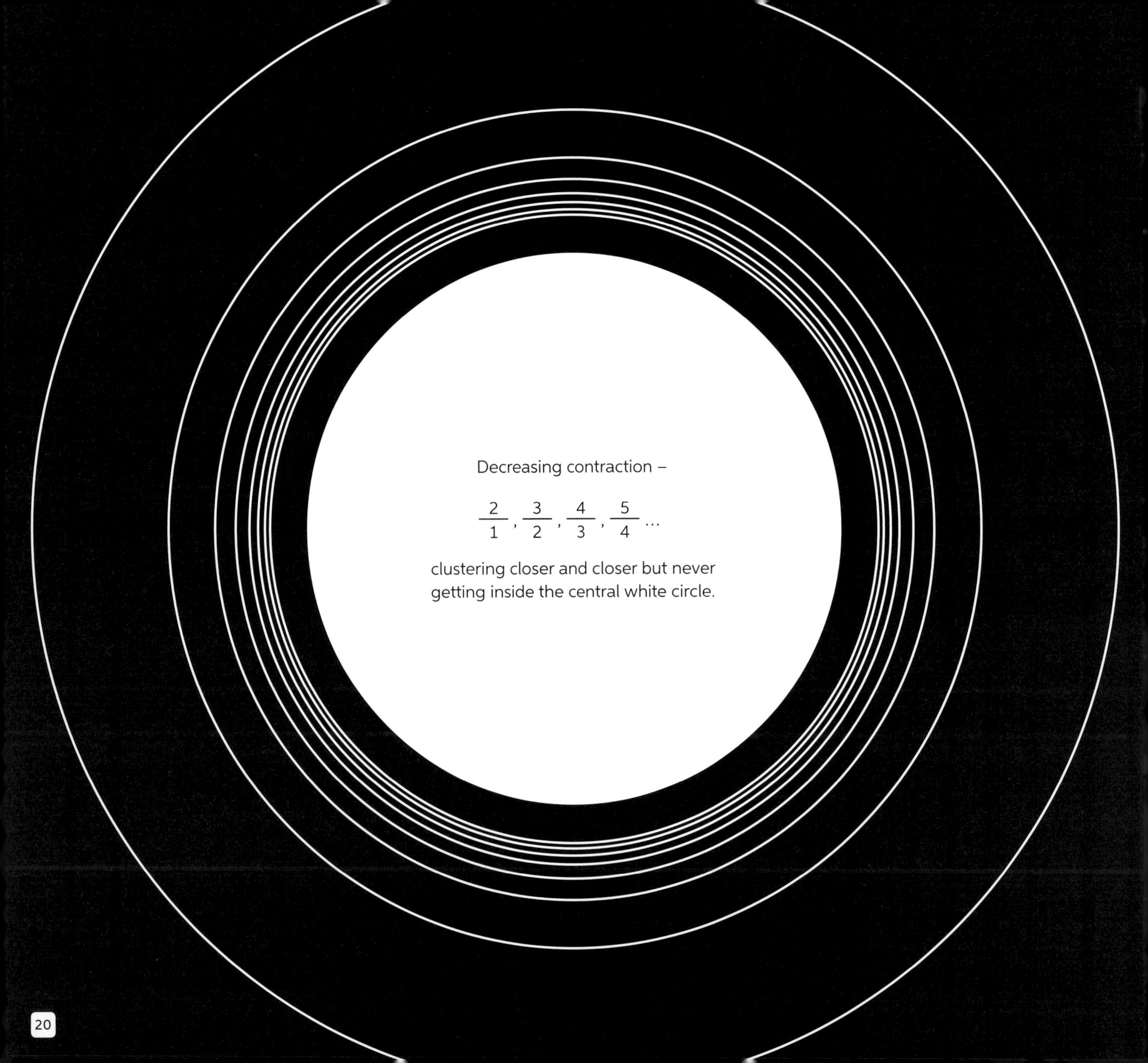

Decreasing contraction –

$$\frac{2}{1}, \frac{3}{2}, \frac{4}{3}, \frac{5}{4} \ldots$$

clustering closer and closer but never getting inside the central white circle.

**That's an interesting badge. What does it stand for—L.O.T.S.?**

Well, the L stands for L.O.T.S., and the 0 for Of, T for The and S for Same. So it stands for L.O.T.S. Of The Same.

**Lots of the same?**

No, not exactly. Not 'Lots Of The Same' but 'L.O.T.S. Of The Same'. And you see that means 'L.O.T.S. Of The Same Of The Same'.

**Oh, I see. And I suppose *that* means 'L.O.T.S. Of The Same Of The Same Of The Same'?**

That's right. And that means 'L.O.T.S. Of The Same Of The Same Of The Same Of The Same'.

**And *that* means 'L.O.T.S. Of The Same Of The Same Of The Same Of The Same Of The Same'.**

Yes. And that means 'L.
Same Of The Same'.

**Which means "L.O.T.S. O
Of The Same Of The Sam**

This conversation is proceeding (new developments will be reported in our next edition).

This machine produces a series of numbers called a **Geometric Progression** (GP).

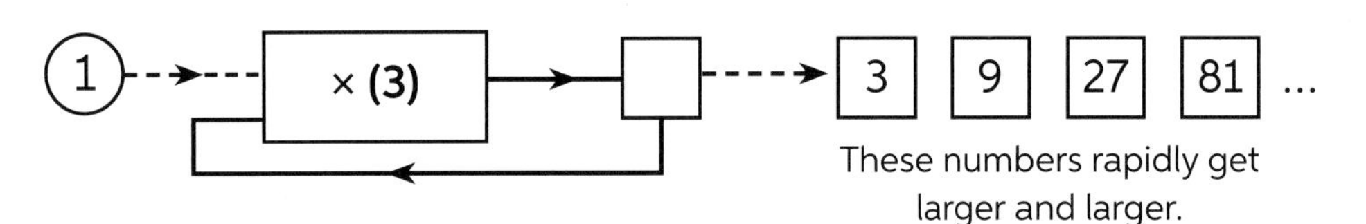


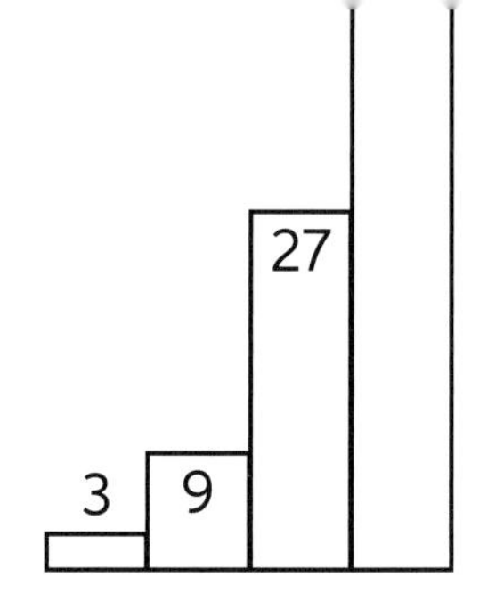


This machine produces a series of numbers called an **Arithmetic Progression** (AP).

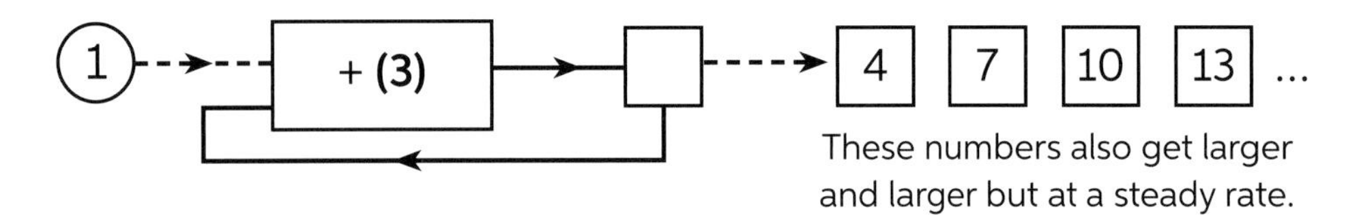


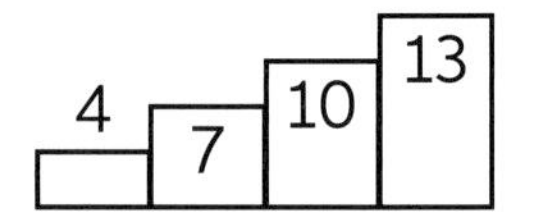


This machine also produces a GP but because the multiplying factor is less than 1, the numbers get rapidly smaller.

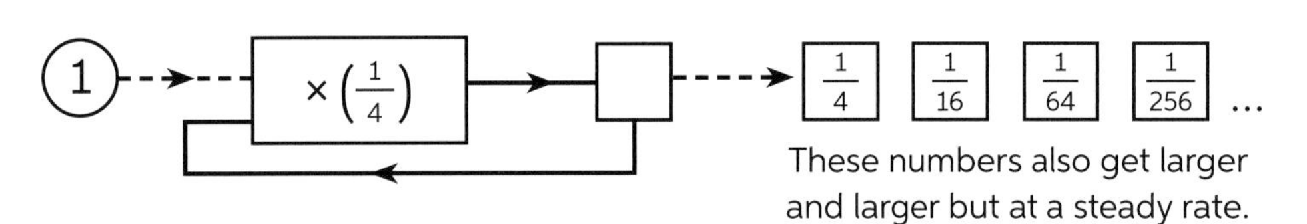


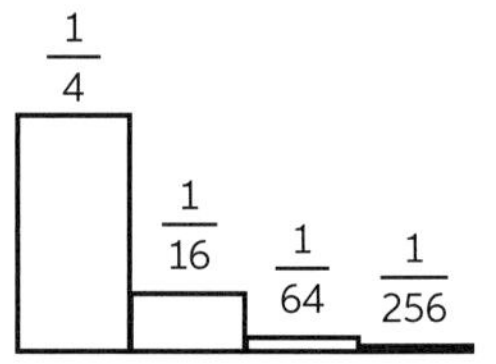


We now introduce some new operators into these machines:

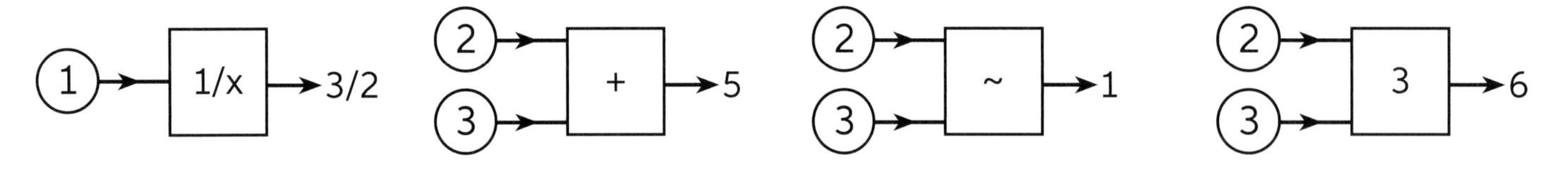


The reader is invited to work out what number series these machines will generate:

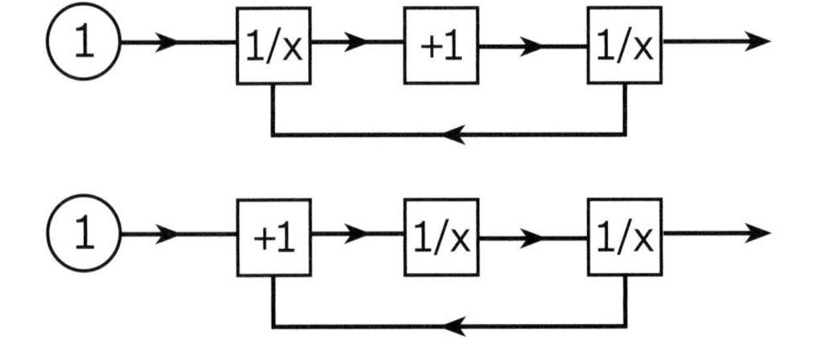


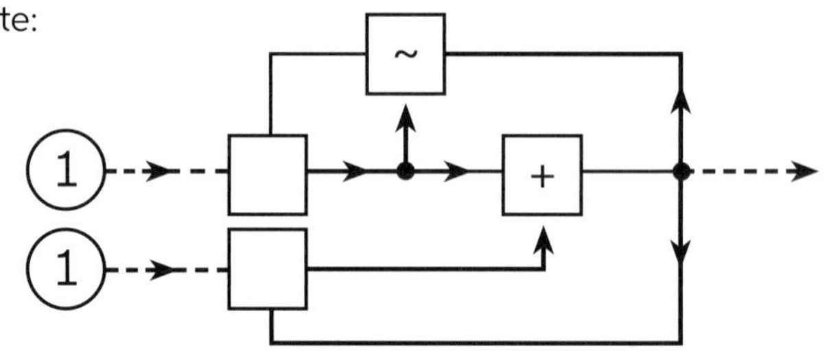


… and then to invent some new machines.

Machines for the recursive writing of **LOTS**

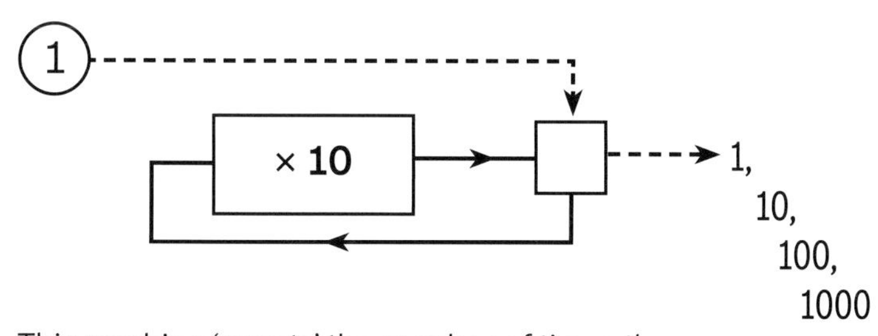


This machine 'counts' the number of times the letter L appears at successive stages.

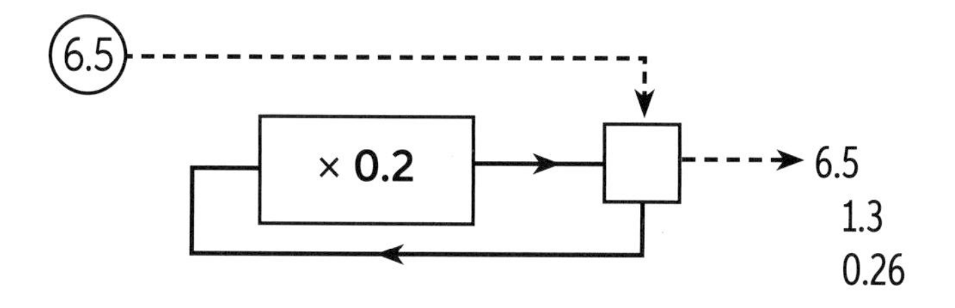


This machine gives (approximately) the size of the letters in successive stages.

In the largest letters (stage 1) the word LOTS is written once

Stage 2 writes the word LOTS 10 times. So LOTS appears 1 + 10 times

Stage 3 writes the word LOTS 100 times. So LOTS appears 1 + 10 + 100 times

Stage 4 writes the word LOTS 1000 times.

This is the stage reached above, where LOTS appears 1111 times.

Here is a machine to do this calculation:

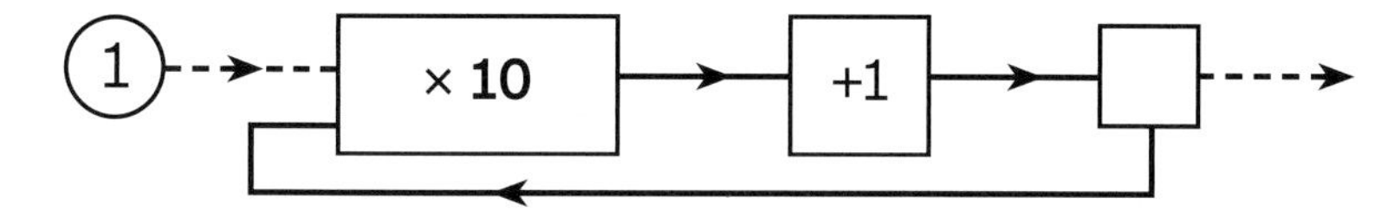

# Magic Staircases

Imagine a staircase that goes up and up ····
···· for ever?
···· there is always a next step.

Going down – of course most staircases stop at ground level.
But magic staircases are different –
For one thing, the steps are not all the same size.
It might be that each step takes you down
only half the distance of the one above it.

You might suppose in that case that after a while you
would start going down two steps, or four steps,
eight steps ··· at a time.

But no – because it is a part of the magic that
every time you go down a step you shrink to
half your size. [Of course it has to be like that,
because otherwise your feet would be too big to fit
on the steps.] So it always seems as though the steps
stay the same size.

To build a magic staircase
they have to start at the top and work downwards.
They have to make some careful calculations
because if they started too high they might never reach the ground.
Builders have made this mistake before now – there's one in Arizona for instance –
they're still building it although they have got 177000 000 000 000 000 000 000 000 000 000 000
000 000 000 000 000 000 000 ...

## The Algebra of Magic Staircases

Not all magic staircases reduce the steps by ½, some reduce by ⅓, some by ¾, some by 0.9 . . .

So we'll call this reduction factor by the letter R.

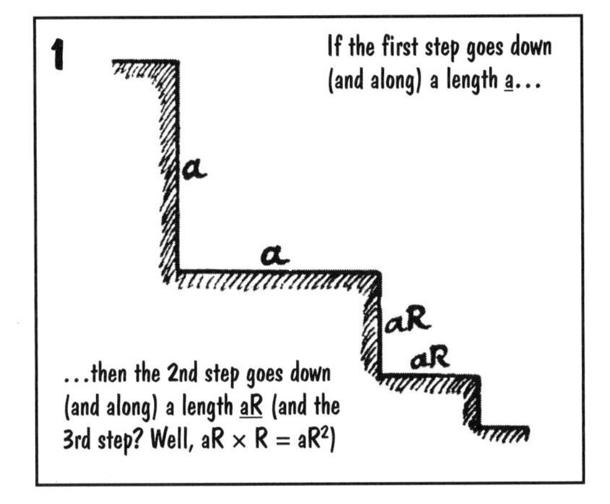

1
If the first step goes down (and along) a length a...
a
a
aR
aR
...then the 2nd step goes down (and along) a length aR (and the 3rd step? Well, aR × R = aR²)


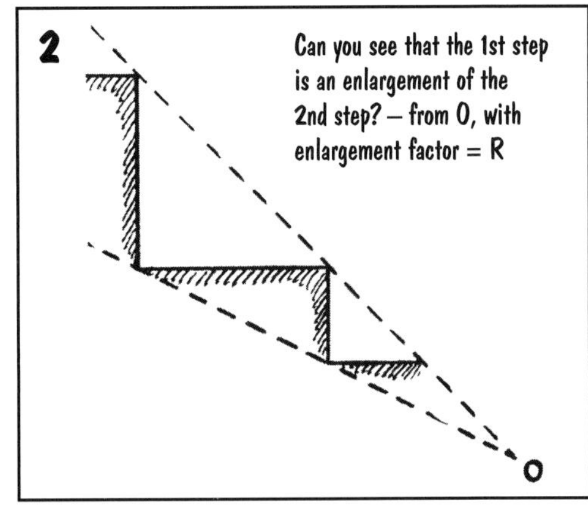

2
Can you see that the 1st step is an enlargement of the 2nd step? – from O, with enlargement factor = R
O


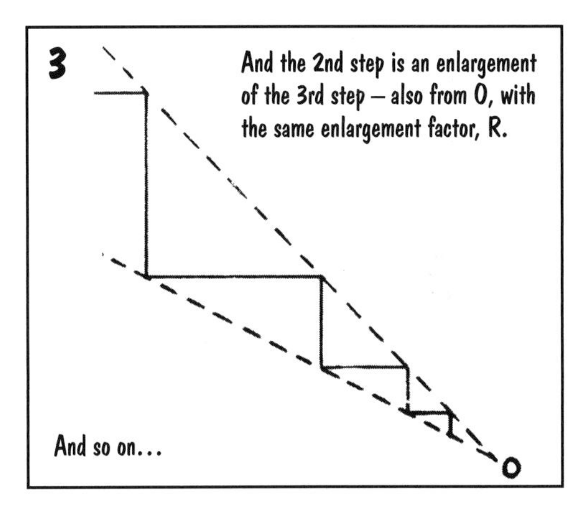

3
And the 2nd step is an enlargement of the 3rd step – also from O, with the same enlargement factor, R.
And so on...
O


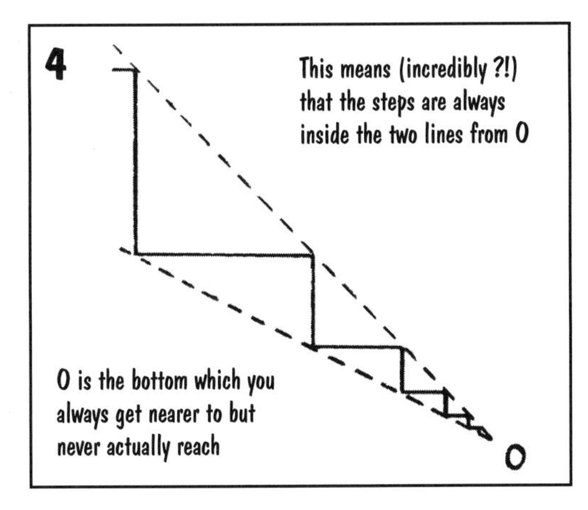

4
This means (incredibly ?!) that the steps are always inside the two lines from O
O is the bottom which you always get nearer to but never actually reach
O


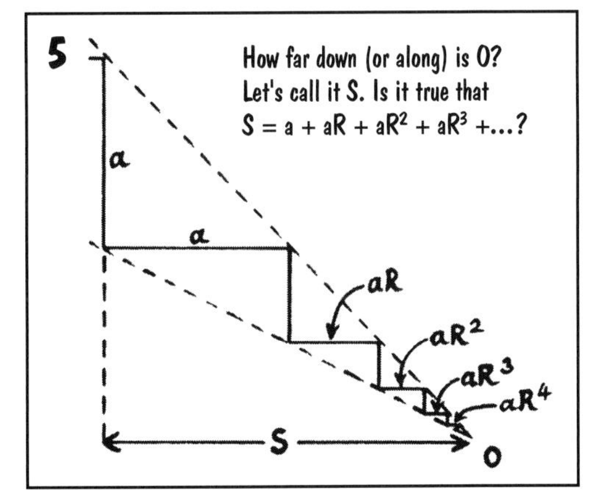

5
How far down (or along) is O? Let's call it S. Is it true that S = a + aR + aR² + aR³ +...?
a
a
aR
aR²
aR³
aR⁴
S
O


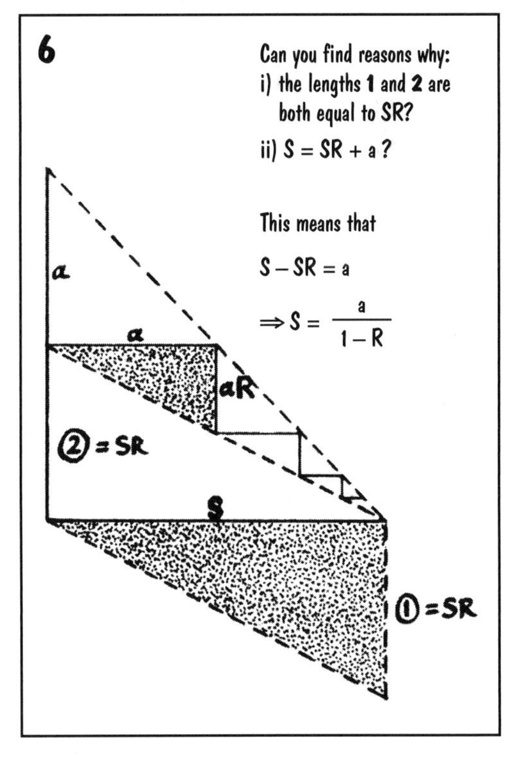

6
Can you find reasons why:
i) the lengths 1 and 2 are both equal to SR?
ii) S = SR + a ?
This means that
S – SR = a
⇒ S = a / 1 – R
a
a
aR
② = SR
S
① = SR


The steps of a magic staircase form a GP: a; aR; aR²; … (R<1)

So you go down (and forward) a distance a + aR + aR² + …

The more steps you take, the closer this distance gets to S.

The algebra shows $S = \dfrac{a}{1 - R}$

In what sense can we write S == a + aR + aR² + …?

There are nearly 400 dots in this box

Here there are about 600 dots …

… nearly 1000

… nearly 1300

… nearly 1600

… nearly 2000

… nearly 2300

… nearly 4500

The dots in this box have become so small you probably can't see them. How many do you think there are?

## A conversation

A: Think of adding up the whole numbers.

B: Like 1 + 2 = 3?

A: Go on.

B: 1 + 2 + 3 = 6. 1 + 2 + 3 + 4 = 10. 1 + 2 +

A: That'll do for now. Notice that the total is always bigger than the last number you add on. Do you agree?

B: Yes, of course—because there's all the other numbers, before the last, one in the total.

A: Right. Now think of this going on for ever.

B: I'm thinking.

A: Will it still be true that each total is bigger than the last number?

B: Yes, even more true!

A: Even if the last number is as big as you can get?

B: Y-yes.

A: So then the total will be a bigger number than the biggest number you can get.

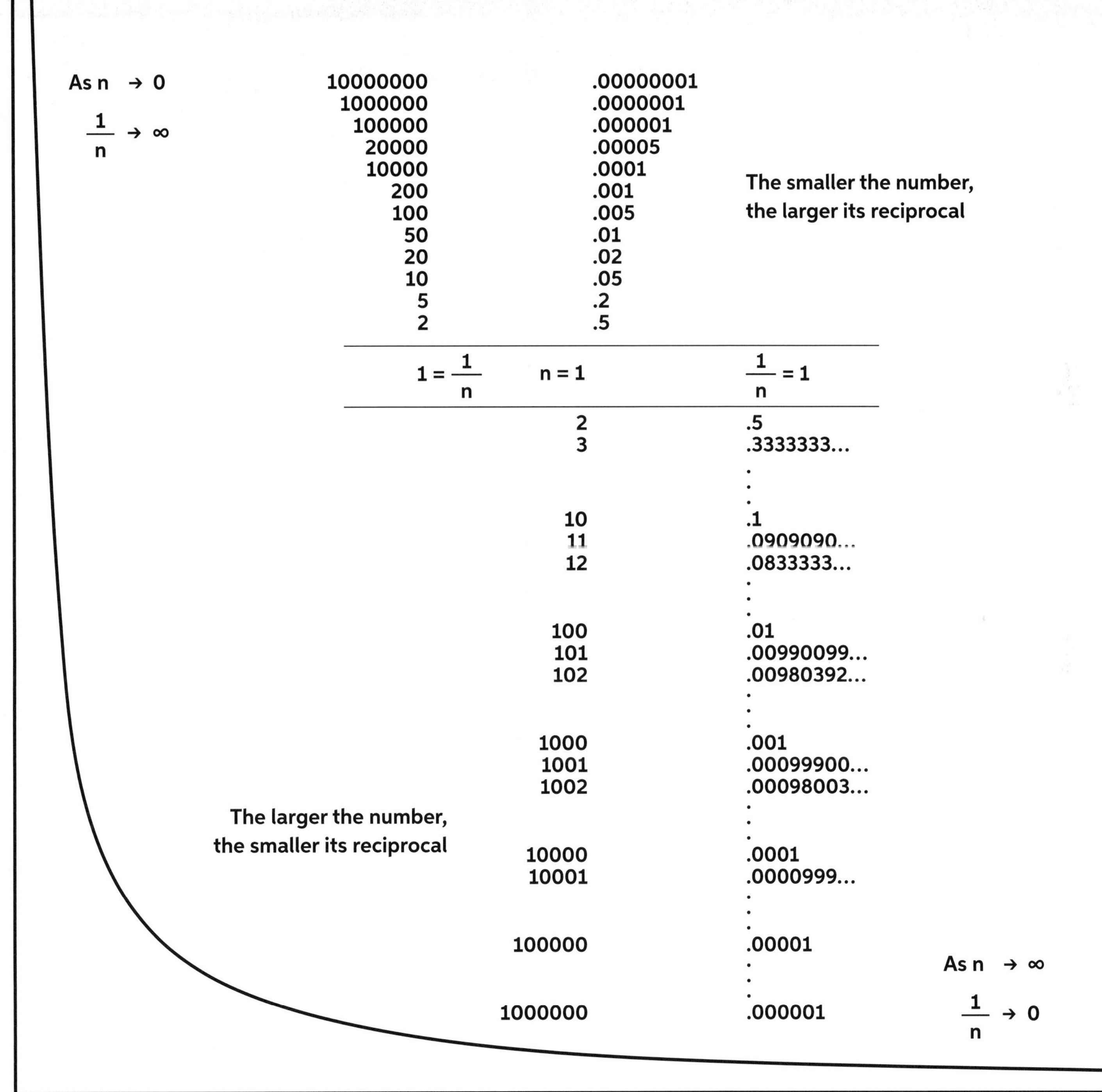
As n → 0
1/n → ∞
10000000 .00000001
1000000 .0000001
100000 .000001
20000 .00005
10000 .0001
200 .001
100 .005
50 .01
20 .02
10 .05
5 .2
2 .5
The smaller the number,
the larger its reciprocal
1 = 1/n
n = 1
1/n = 1
2 .5
3 .3333333...
10 .1
11 .0909090...
12 .0833333...
100 .01
101 .00990099...
102 .00980392...
1000 .001
1001 .00099900...
1002 .00098003...
The larger the number,
the smaller its reciprocal
10000 .0001
10001 .0000999...
100000 .00001
As n → ∞
1/n → 0
1000000 .000001
10000000 .0000001

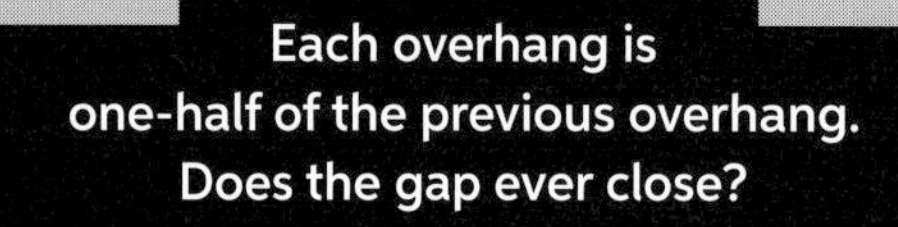

Each overhang is
one-half of the previous overhang.
Does the gap ever close?


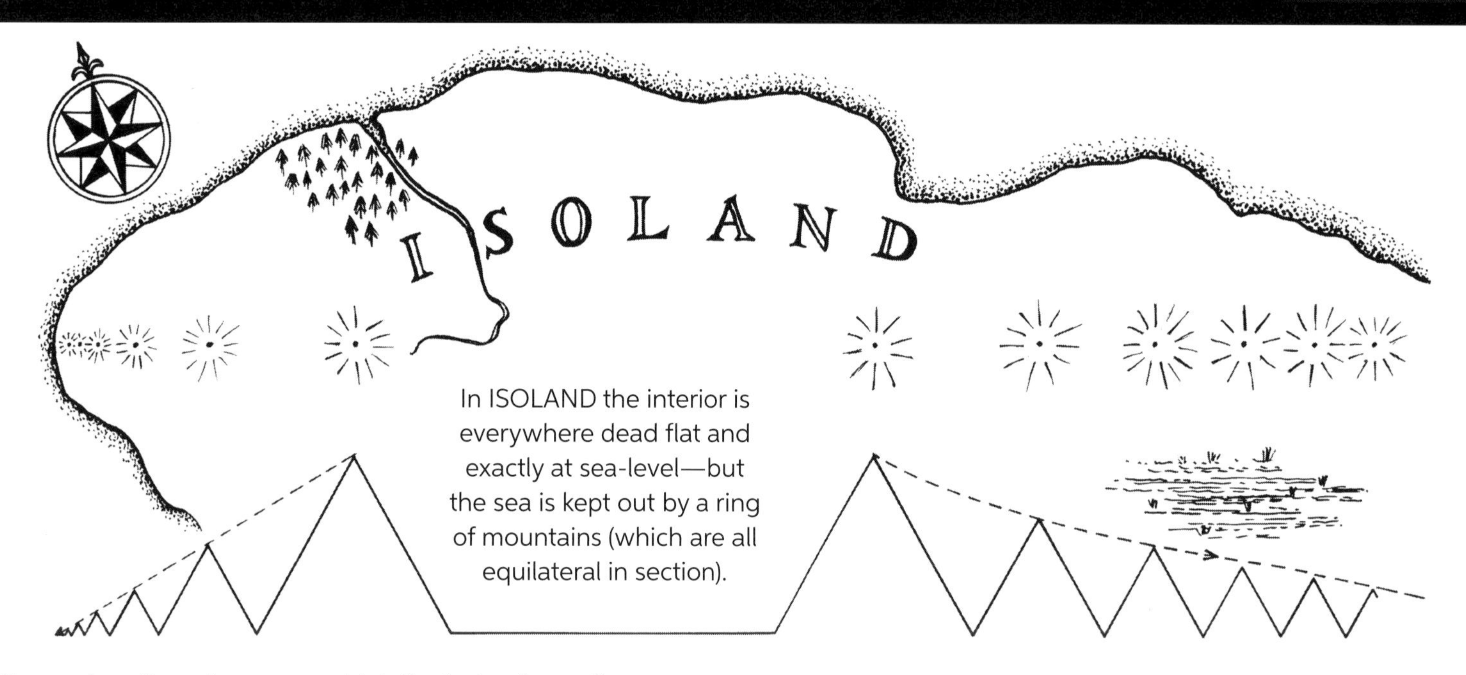

ISOLAND
In ISOLAND the interior is everywhere dead flat and exactly at sea-level—but the sea is kept out by a ring of mountains (which are all equilateral in section).


At one place there is a range which the Isolanders call the Geometric Mountains. The sides of these mountains decrease regularly as they approach the sea—the sloping side of the highest is $1/2$ mile, the next $1/4$ mile, then $1/8$ mile, $1/16$ mile, and so on.

The Isolanders like these mountains because from the top of them you can just see the beach: and it is just a mile-and-a-half walk to get there, up and down the mountains.

At another place there is a range which is called the Harmonic Mountains. These also get smaller regularly—the highest has a side of $1/2$ mile, but the next is $1/3$ mile, then $1/4$ mile, then $1/5$ mile, and so on.

You can never see the beach from these mountains—the others always get in the way. And it's a much longer walk to the beach, much longer, much, much longer.

Will your spiral ever stop?

There is always another line.

What is its total length?

Straighten it out, unwinding it down the vertical line.

Notice as you do this the corners always end up on the dotted line.

So, although there's always more to unwind, the bit of the spiral left is always between the dotted line and the vertical.

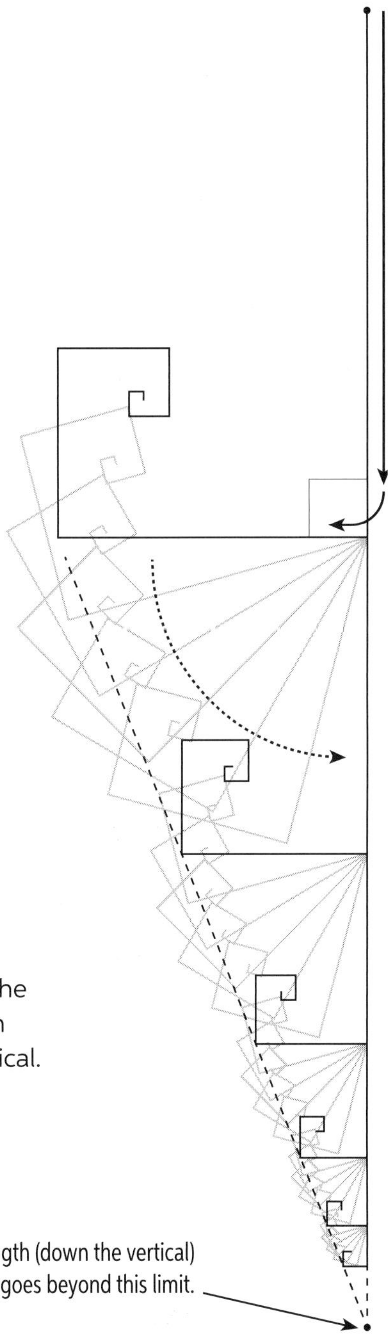

The total length (down the vertical) never goes beyond this limit.

Here is another spiral made to a different rule.

Its length is

$$1 + \frac{1}{2} + \frac{1}{3} + \frac{1}{4} + \frac{1}{5} + \ldots$$

How long is that?

To find out unwind it

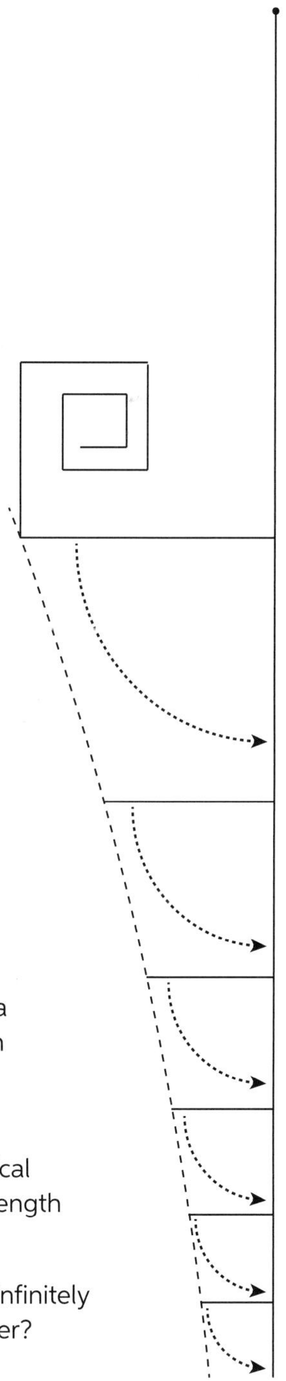

This time the corners are not on a straight line but on a curve which seems to be getting more nearly parallel to the vertical.

If the curve never meets the vertical line, there is no limit to the total length of the spiral.

Can this really be possible?—an infinitely long spiral on this little bit of paper?

# THE BOOK OF SAND

## Jorge Luis Borges

The line is made up of an infinite number of points; the plane of an infinite number of lines; the volume of an infinite number of planes; the hypervolume of an infinite number of volumes . . . No, unquestionably this is not – *more geometrico* – the best way of beginning my story. To claim that it is true is nowadays the convention of every made-up story. Mine, however, *is* true.

I live alone in a fourth-floor apartment on Belgrano Street, in Buenos Aires. Late one evening, a few months back, I heard a knock at my door. I opened it and a stranger stood there. He was a tall man, with nondescript features – or perhaps it was my myopia that made them seem that way. Dressed in grey and carrying a grey suitcase in his hand, he had an unassuming look about him. I saw at once that he was a foreigner. At first, he struck me as old; only later did I realize that I had been misled by his thin blond hair, which was, in a Scandinavian sort of way, almost white. During the course of our conversation, which was not to last an hour, I found out that he came from the Orkneys.

I invited him in, pointed to a chair. He paused awhile before speaking. A kind of gloom emanated from him – as it does now from me.

'I sell Bibles,' he said.

*continued on page 47*

## A Zeno paradox

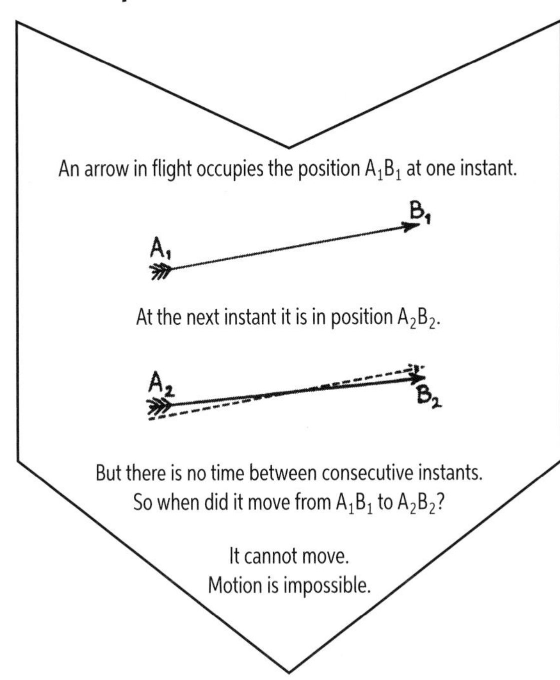


Is a line really made up of points? Can time be broken down into ultimately invisible 'instants'?

Zeno (c. 450BC) produced a series of paradoxes showing that this idea leads to absurd conclusions—An arrow cannot move!

He held the contrary theory—that space and time were continuous. But this also leads to difficulties and the two theories have been debated over many centuries.

## A Galileo paradox

The paradox described by Galileo (1564-1642) is an example of these difficulties:

Any polygon behaves like the hexagons: if there were a thousand sides they would lie along HT, separated by a thousand gaps. These, added up, would be the difference between the perimeters of the two 1000-gons.

A million sides and there is no problem. But a circle…

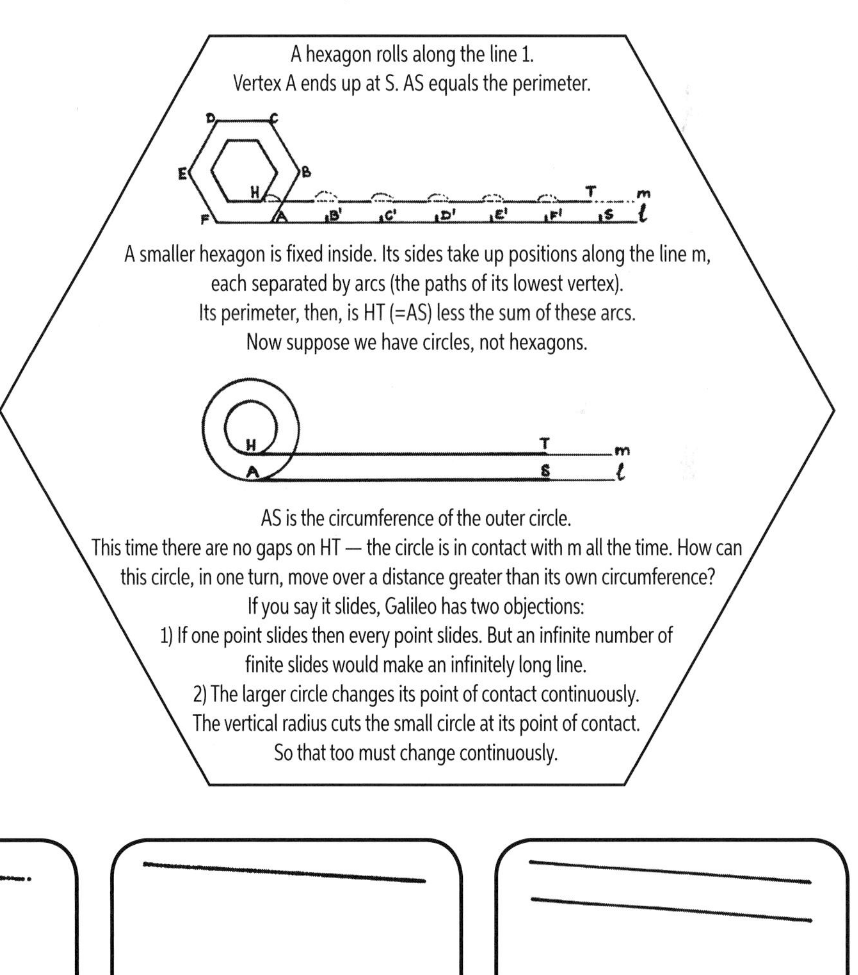

A hexagon rolls along the line 1.
Vertex A ends up at S. AS equals the perimeter.
D
C
E
B
H
T
m
F
A
B'
C'
D'
E'
F'
S
l
A smaller hexagon is fixed inside. Its sides take up positions along the line m, each separated by arcs (the paths of its lowest vertex).
Its perimeter, then, is HT (=AS) less the sum of these arcs.
Now suppose we have circles, not hexagons.
H
T
m
A
S
l
AS is the circumference of the outer circle.
This time there are no gaps on HT — the circle is in contact with m all the time. How can this circle, in one turn, move over a distance greater than its own circumference?
If you say it slides, Galileo has two objections:
1) If one point slides then every point slides. But an infinite number of finite slides would make an infinitely long line.
2) The larger circle changes its point of contact continuously.
The vertical radius cuts the small circle at its point of contact.
So that too must change continuously.

- Mathematicians in the 16th and 17th centuries made a lot of progress in finding areas by thinking of them as made up of an infinite number of straight lines. There are problems in this idea: what is the area of a line? How can you add up an infinite number of things? …
- Here is the way one man, Cavalieri (1598-1647), tackled this problem for a parallelogram. He used the lines to find the *ratio* of areas of two similar parallelograms.

The dotted lines which fill the areas are all of length a in A and p in P.

So ratio of areas A : P is
(a + a + a + …): (p + p + p + …)

But how many a's and how many p's?
Cavalieri said this would depend on the lengths of the other sides—b and q.

So the ratio of areas A : P

$$= (b \times a) : (q \times p) = \frac{b}{q} \times \frac{a}{p}$$

But also, because the parallelograms are similar,

$$\frac{a}{p} = \frac{b}{q}$$

So the ratio of areas A : P

$$= \frac{a}{p} \times \frac{b}{q} = a^2 : p^2$$

- This is a correct result which no one argued with.

But many people at the time could, and did, argue with the reasoning. One objection, for instance, pointed to the kind of fallacy that can be 'proved' by summing lines to find areas:

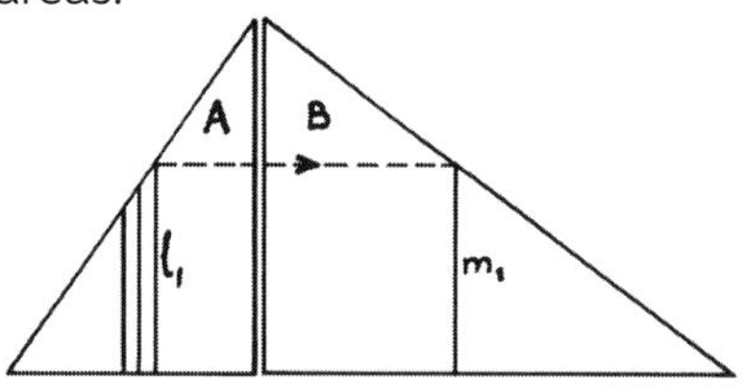


Look at the two right-angled triangles, areas A and B. Suppose A is made up of lines like l: that is, $A = l_1 + l_2 + l_3 + \ldots$

The dotted line shows a way of finding a line $m_1$ in triangle B which is equal to $1_1$.

Every line l, in A has an equal partner, m, in B.

And vice versa.

So $l_1 + l_2 + l_3 + \ldots m_1 + m_2 + m_3 \ldots$
So A=B

The two triangles have the same area!

- Nevertheless, Cavalieri went on to develop the technique. He showed, for instance, that his result for similar parallelograms is also true for similar figures drawn within such parallelograms.

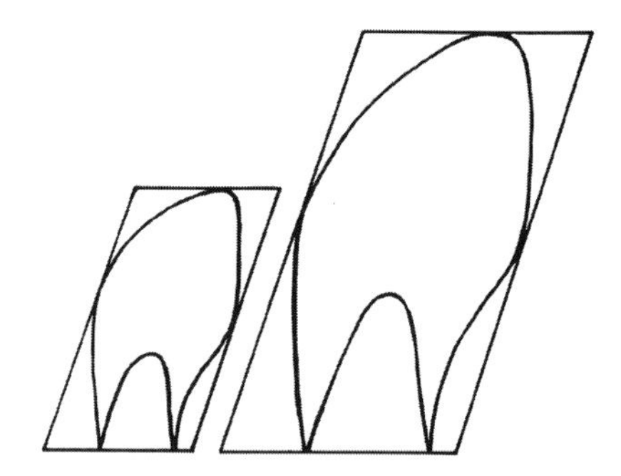

This enabled him to find formulae for areas bounded by such curves as

$y = x^2$
$y = x^5 + x^3$
etc.

Was it all based on a fallacy?

Mathematicians have often developed effective techniques which were based on very shaky arguments.

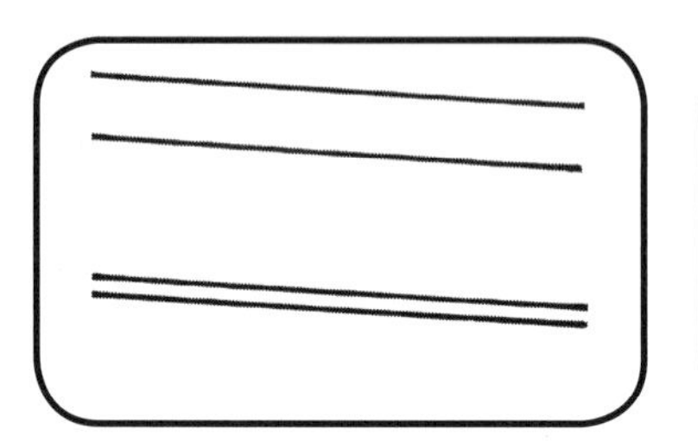

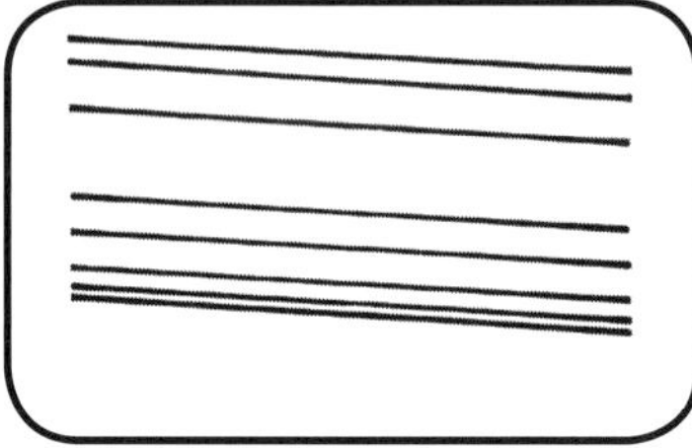

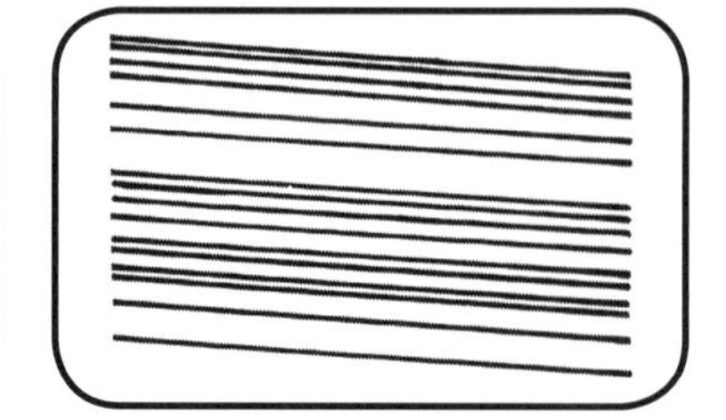

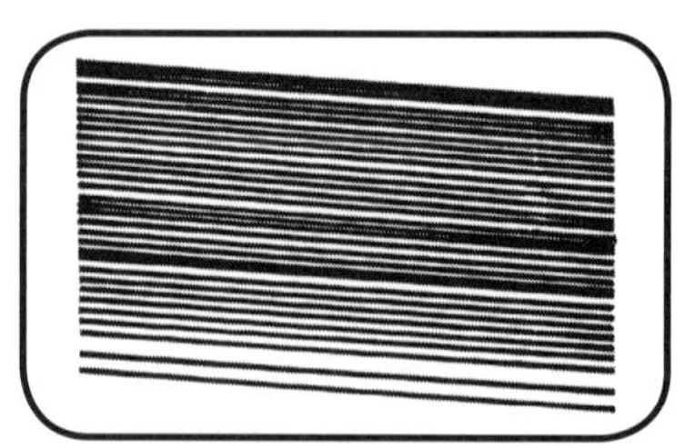

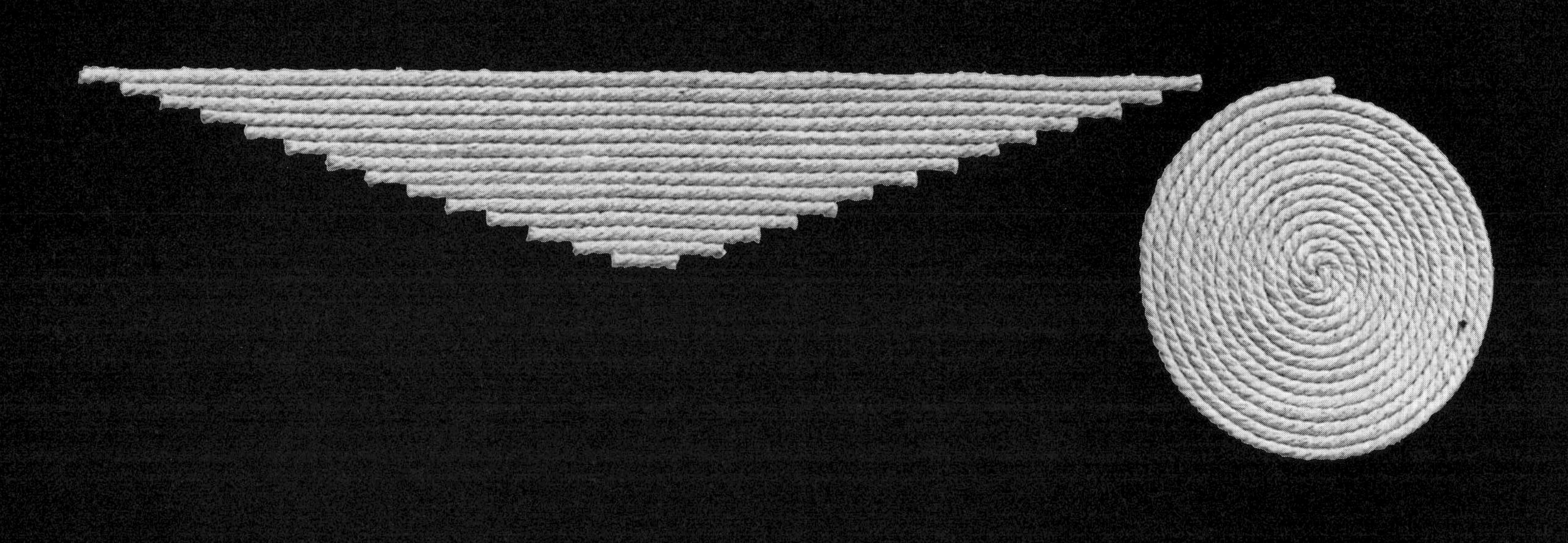

‘Circles’ of string become ‘triangles’.

Base of ‘triangle’ — circumference of ‘circle’.

Height of ‘triangle’ — radius of ‘circle’.

So area of circle = $^1/_2 \times$ (circumference) $\times$ (radius).

The thinner the string, the more nearly are they true circles and triangles.

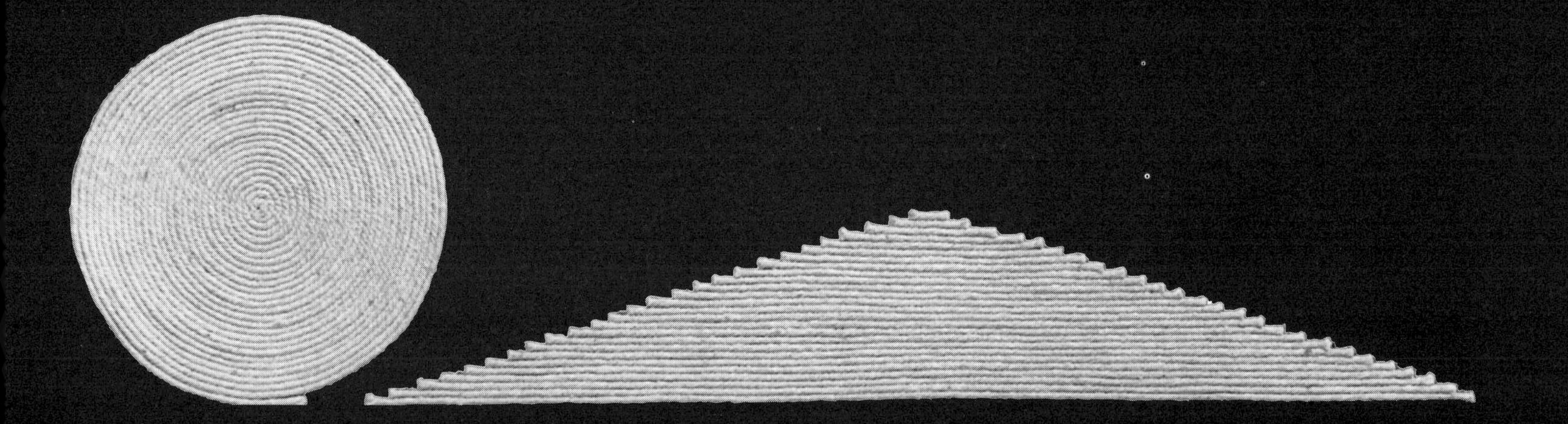

Two ways in which a computer might try to draw a circle

**1 As a set of dots:**

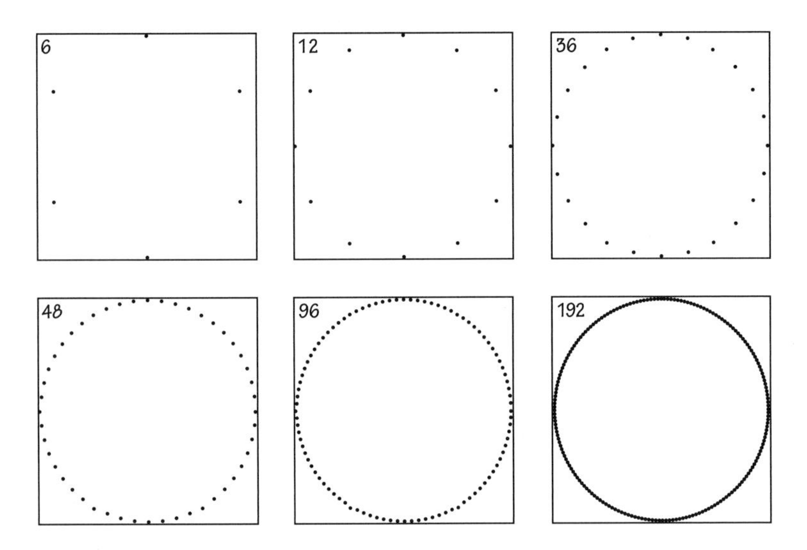


**2 As a set of lines:**

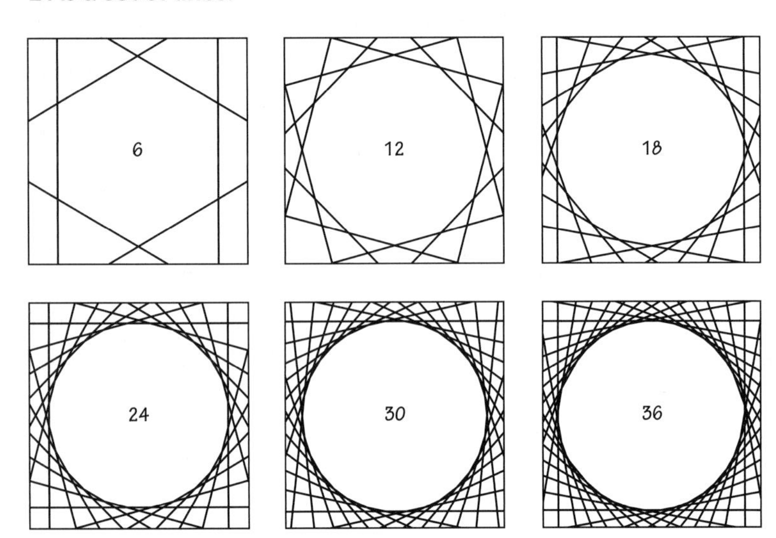


**Kepler (1571-1630) wrote:**

'The circumference of a circle has as many parts as points, namely an infinite number; each of these can be regarded as the base of an isosceles triangle with equal sides, so that there are an infinite number of triangles in the area of a circle'.

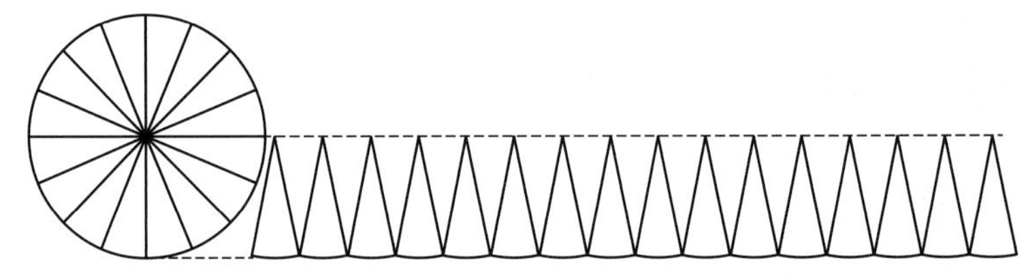

Allowing for the gaps, the area of these 'traingles' make half the area of a rectangle, i.e. the area of a right triangle.

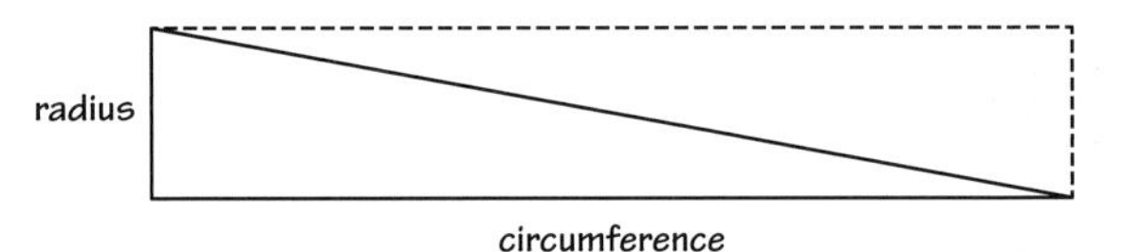


$$\text{Area} = \frac{1}{2}\ (\text{circumference}) \times (\text{radius})$$

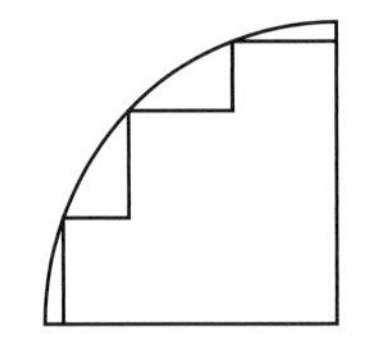

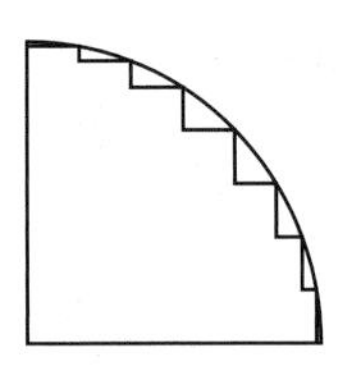

A quadrant of a circle seems to have approximately the same length as the total up-and-along parts of a staircase. The more steps you take the more they get like the circle.

But the total 'ups' = radius and total 'alongs' = radius.

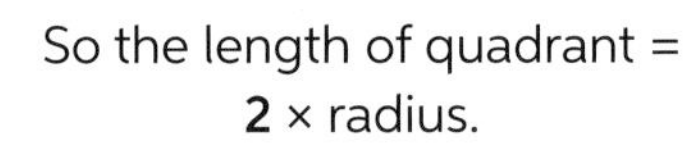

So the length of quadrant = **2** × radius.

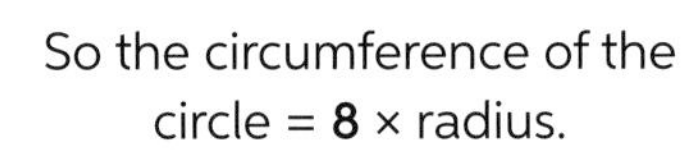

So the circumference of the circle = **8** × radius.

This means π = 4.

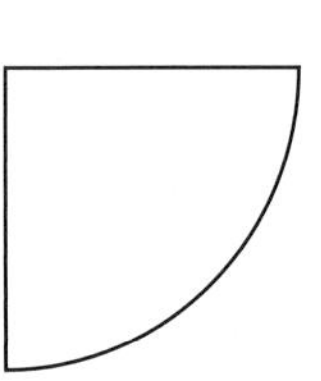

if a straight line
is an infinite circle
then where is its centre?

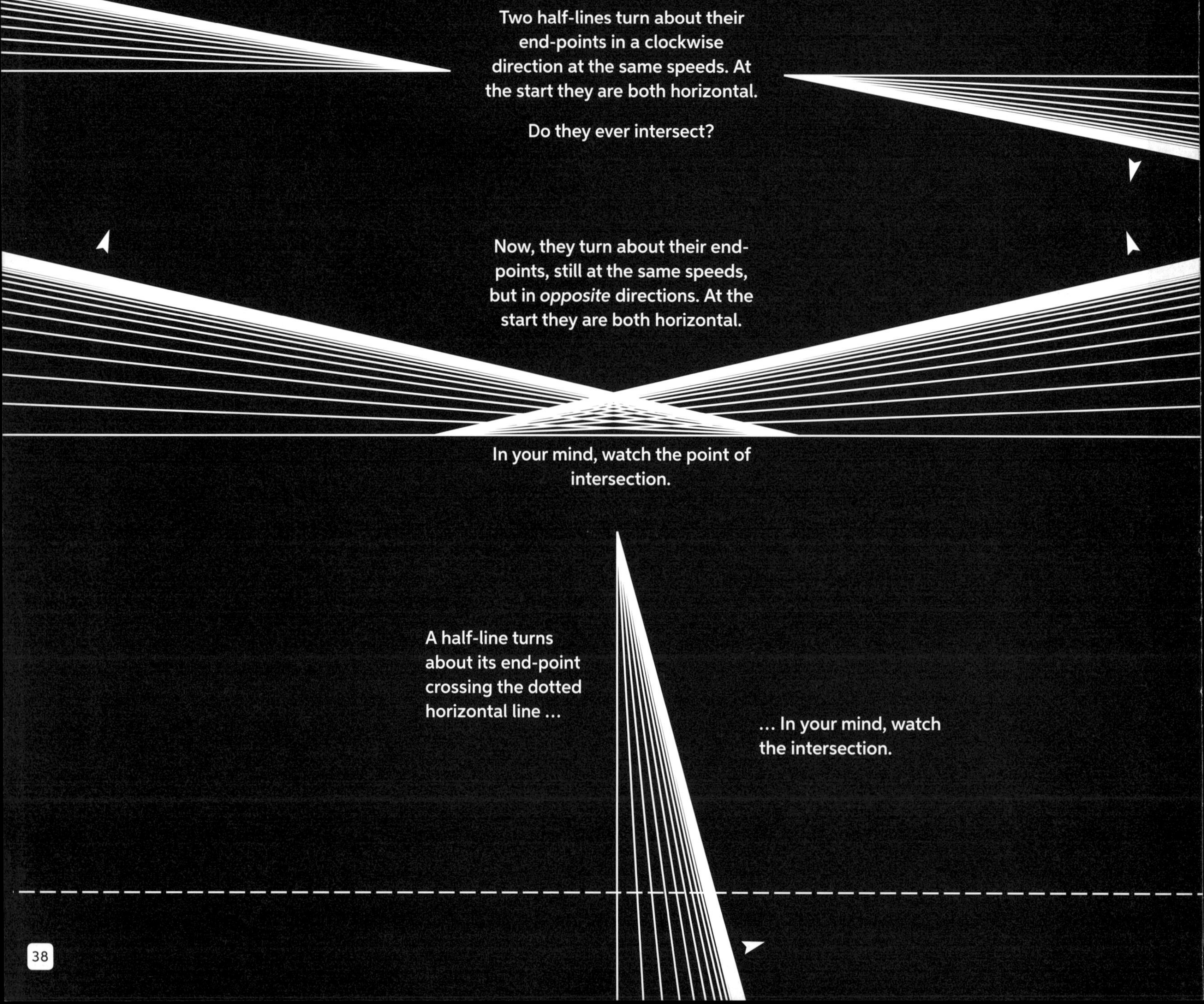

Two half-lines turn about their end-points in a clockwise direction at the same speeds. At the start they are both horizontal.

Do they ever intersect?

Now, they turn about their end-points, still at the same speeds, but in *opposite* directions. At the start they are both horizontal.

In your mind, watch the point of intersection.

A half-line turns about its end-point crossing the dotted horizontal line …

… In your mind, watch the intersection.

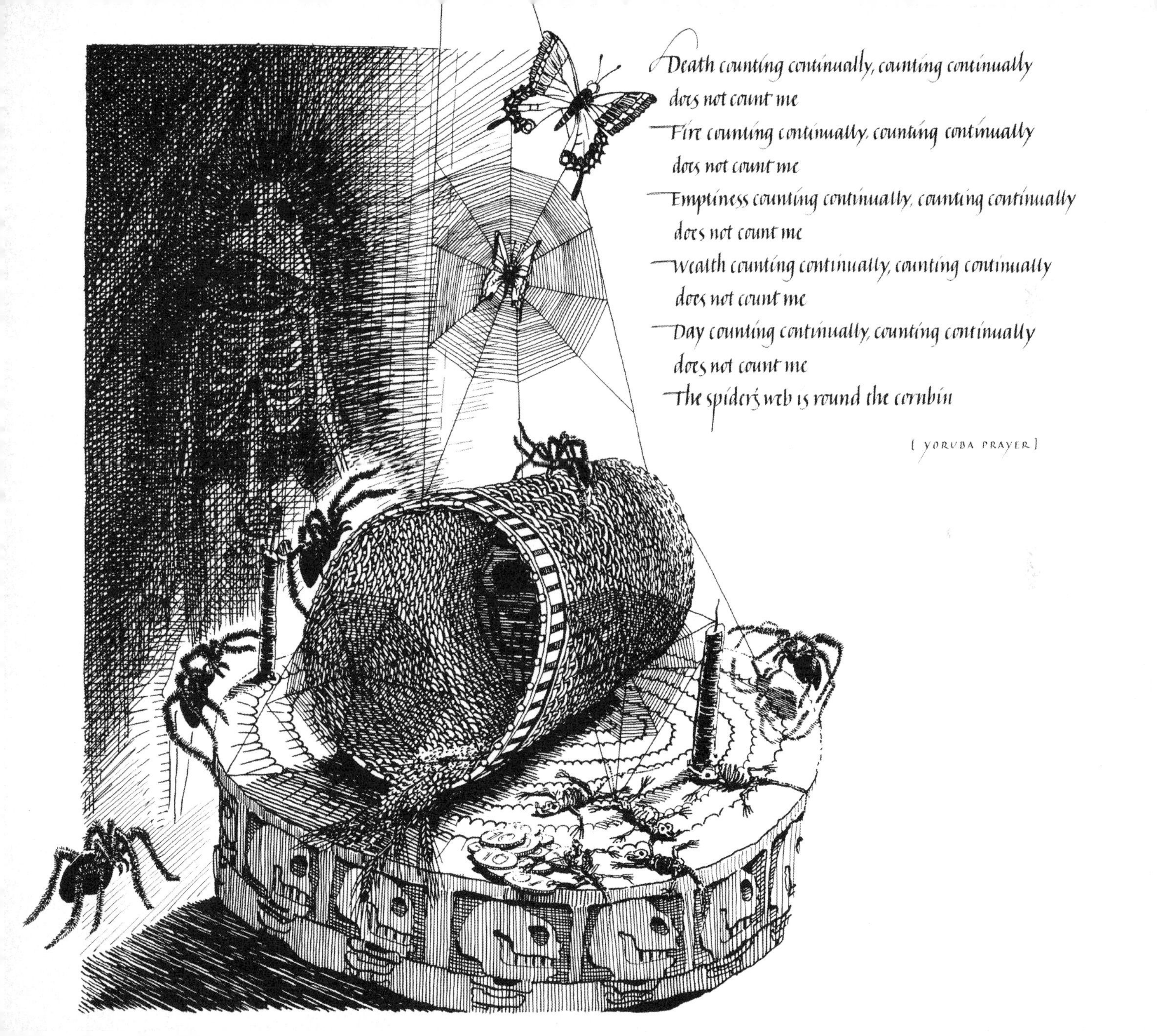
Death counting continually, counting continually
does not count me
Fire counting continually, counting continually
does not count me
Emptiness counting continually, counting continually
does not count me
Wealth counting continually, counting continually
does not count me
Day counting continually, counting continually
does not count me
The spider's web is round the cornbin
[ YORUBA PRAYER ]

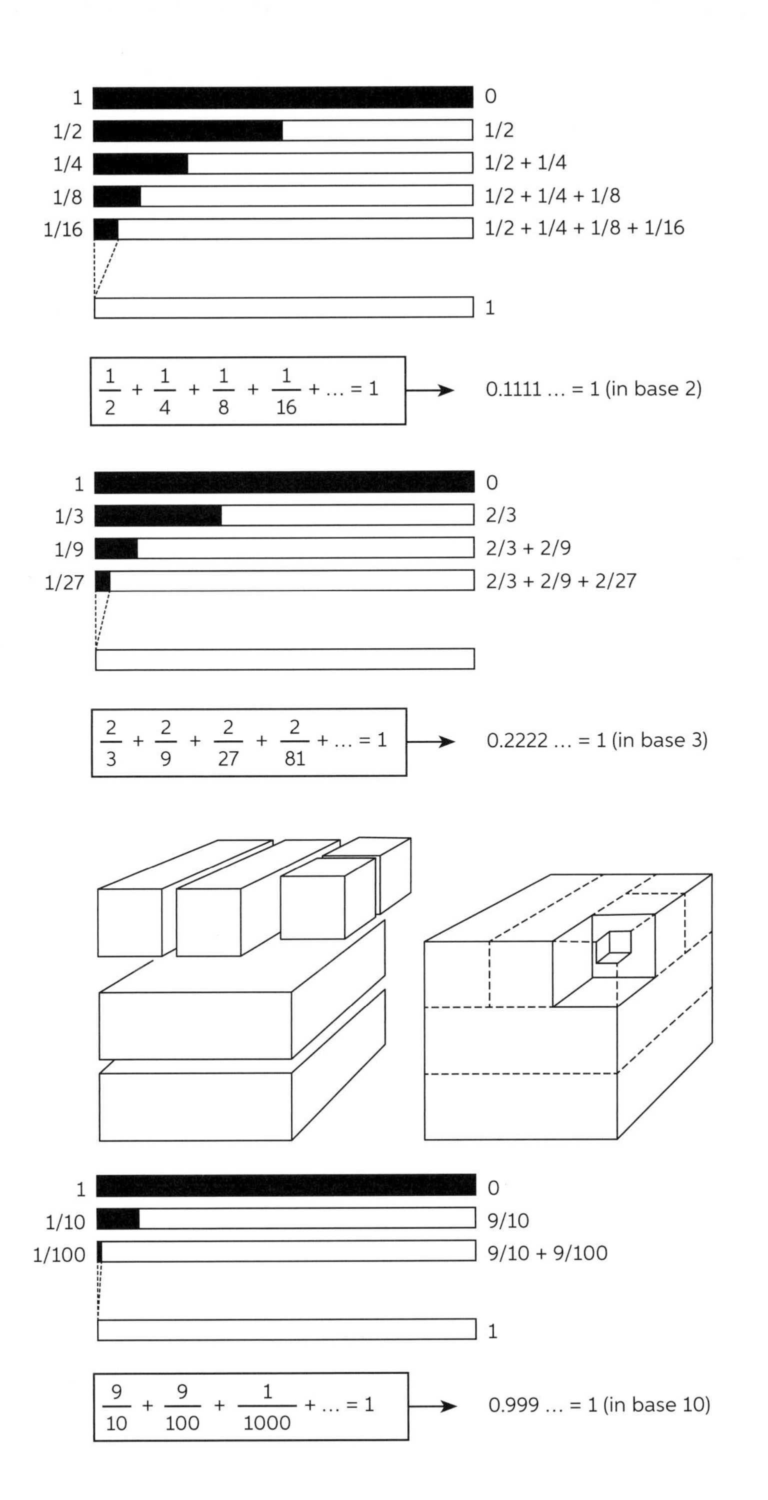


The stranger asked me to find the first page.

I laid my left hand on the cover and, trying to put my thumb on the flyleaf, I opened the book. It was useless. Every time I tried, a number of pages came between the cover and my thumb. It was as it they kept growing from the book.

'Now find the last page.'

Again I failed. In a voice that was not mine, I barely managed to stammer, 'This can't be.'

Still speaking in a low voice, the stranger said, 'It can't be, but it *is*. The number of the pages in this book is no more or less than infinite. None is the first page, none the last. I don't know why they're numbered in this arbitrary way. Perhaps to suggest that the terms of an infinite series admit any number.

Then, as if he were thinking aloud, he said, 'If space is infinite, we may be at any point in space. If time is infinite, we may be at any point in time.'

His speculations irritated me. 'You are religious, no doubt?' I asked him.

'Yes, I'm a Presbyterian. My conscience is clear. I am reasonably sure of not having cheated the native when I gave him the Word of God in exchange for his devilish book.'

'I assured him that he had nothing to reproach himself for, and I asked if he were just passing through this part of the world. He replied that he planned to return to his country in a few days. It was then that I learned that he was a Scot from the Orkney Islands. I told him I had a great personal affection for Scotland, through my love of Stevenson and Hume.

'You mean Stevenson and Robbie Burns,' he corrected.

While we spoke, I kept exploring the infinite book. With feigned indifference, I asked, 'Do you intend to offer this curiosity to the British Museum?'

'No. I'm offering it to you,' he said, and he stipulated a rather high sum for the book.

I answered, in all truthfulness, that such a sum was out of my reach, and I began thinking. After a minute or two, I came up with a scheme.

'I propose a swap,' I said. 'You got this book for a handful of rupees and a copy of the Bible. I'll offer you the amount of my pension cheque, which I've just collected, and my black-letter Wiclif Bible. I inherited it from my ancestors.'

'A black-letter Wiclif!' he murmured.

I went to my bedroom and brought him the money and the book. He turned the leaves and studied the title page with all the fervour of a true bibliophile.

'It's a deal,' he said.

*continued on page 71*

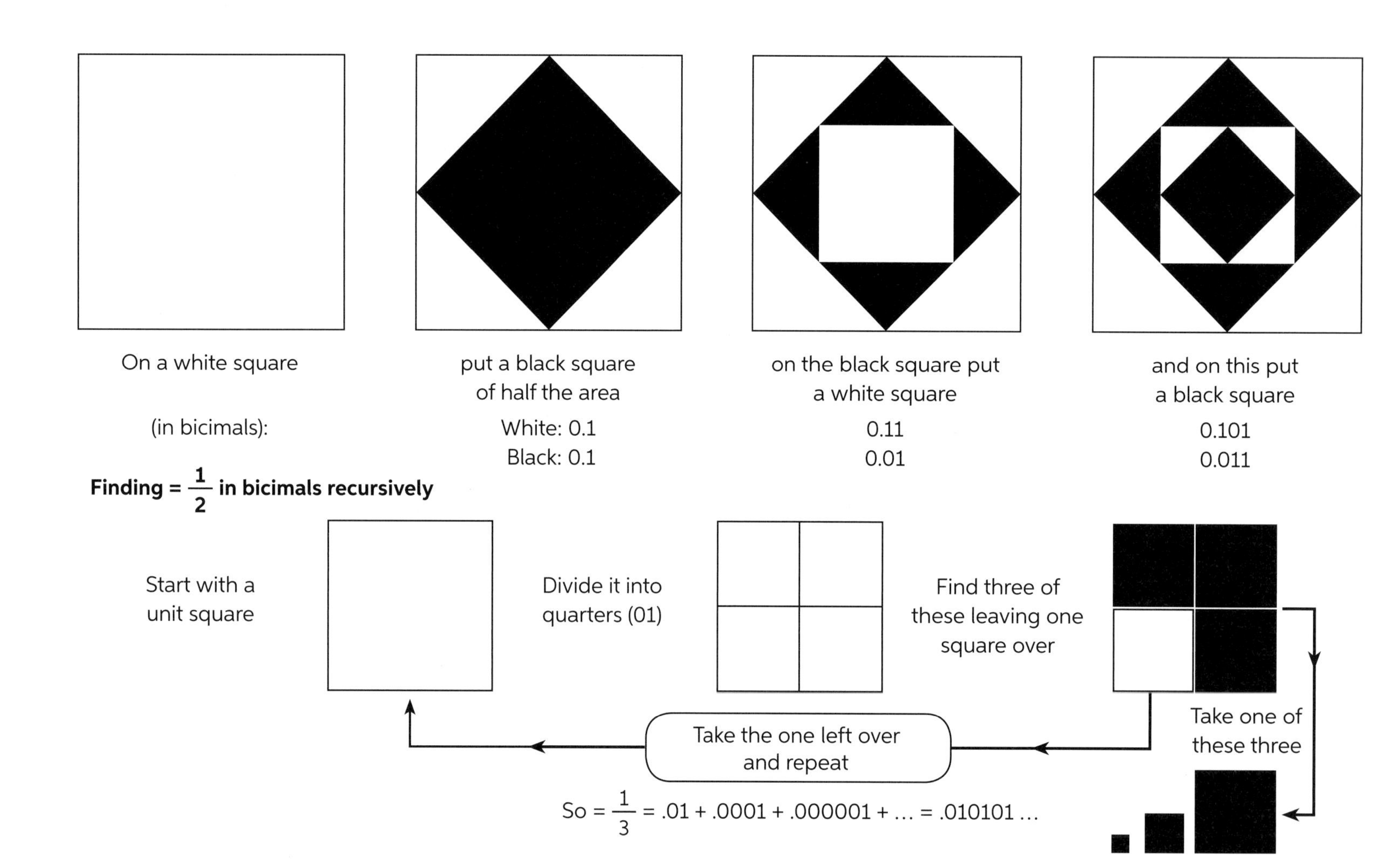
On a white square
put a black square
of half the area
on the black square put
a white square
and on this put
a black square
(in bicimals):
White: 0.1
Black: 0.1
0.11
0.01
0.101
0.011
Finding = $\frac{1}{2}$ in bicimals recursively
Start with a
unit square
Divide it into
quarters (01)
Find three of
these leaving one
square over
Take one of
these three
Take the one left over
and repeat
So = $\frac{1}{3}$ = .01 + .0001 + .000001 + … = .010101 …

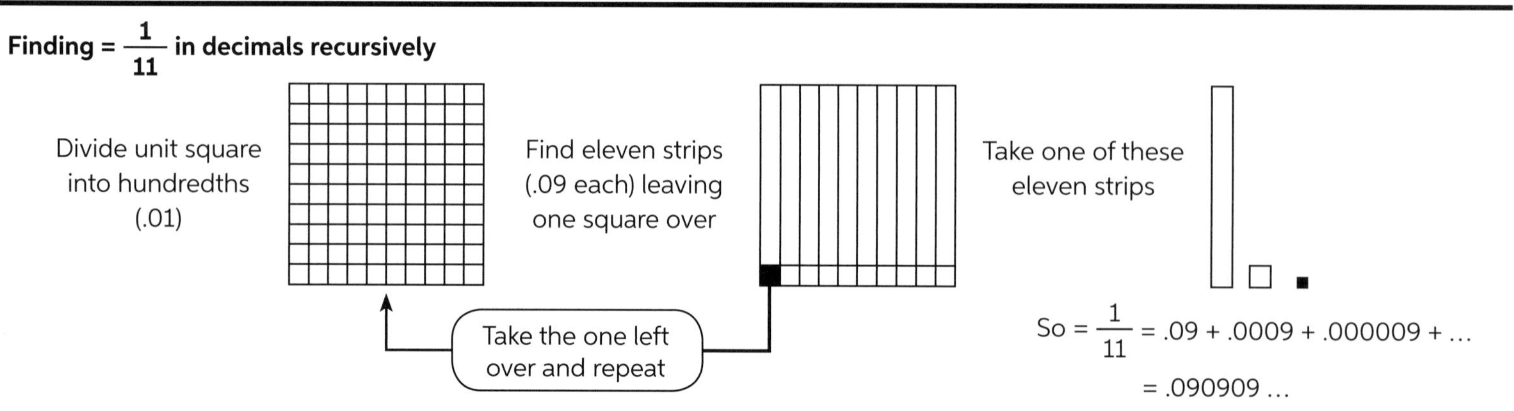
Finding = $\frac{1}{11}$ in decimals recursively
Divide unit square
into hundredths
(.01)
Find eleven strips
(.09 each) leaving
one square over
Take one of these
eleven strips
Take the one left
over and repeat
So = $\frac{1}{11}$ = .09 + .0009 + .000009 + …
= .090909 …

and then a white square

0.1011
0.10101

and then another black square

0.10101
0.01011

—and if you keep doing this (for ever?) the area of white will be 0.1010101 … .0101 … & the area of black will be 0.010101 … 0101…

Now take all the black pieces in the last drawing and put them into the white square in a different way:

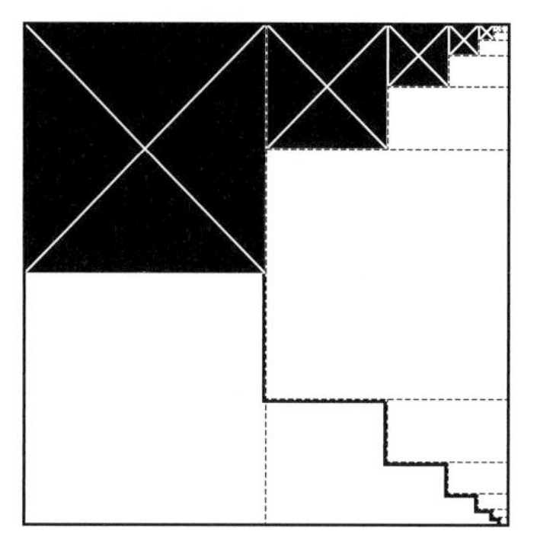

Can you see that the black pieces occupy very nearly 1/3 of the square?

So 0.0101010101 … = 1/3
and 0.1010101010 … = 2/3

Does this argument hold water?

$$\frac{1}{3} + \frac{2}{3} = 1$$

.101010 … + .010101 = .111111 …

So .111111 … = 1

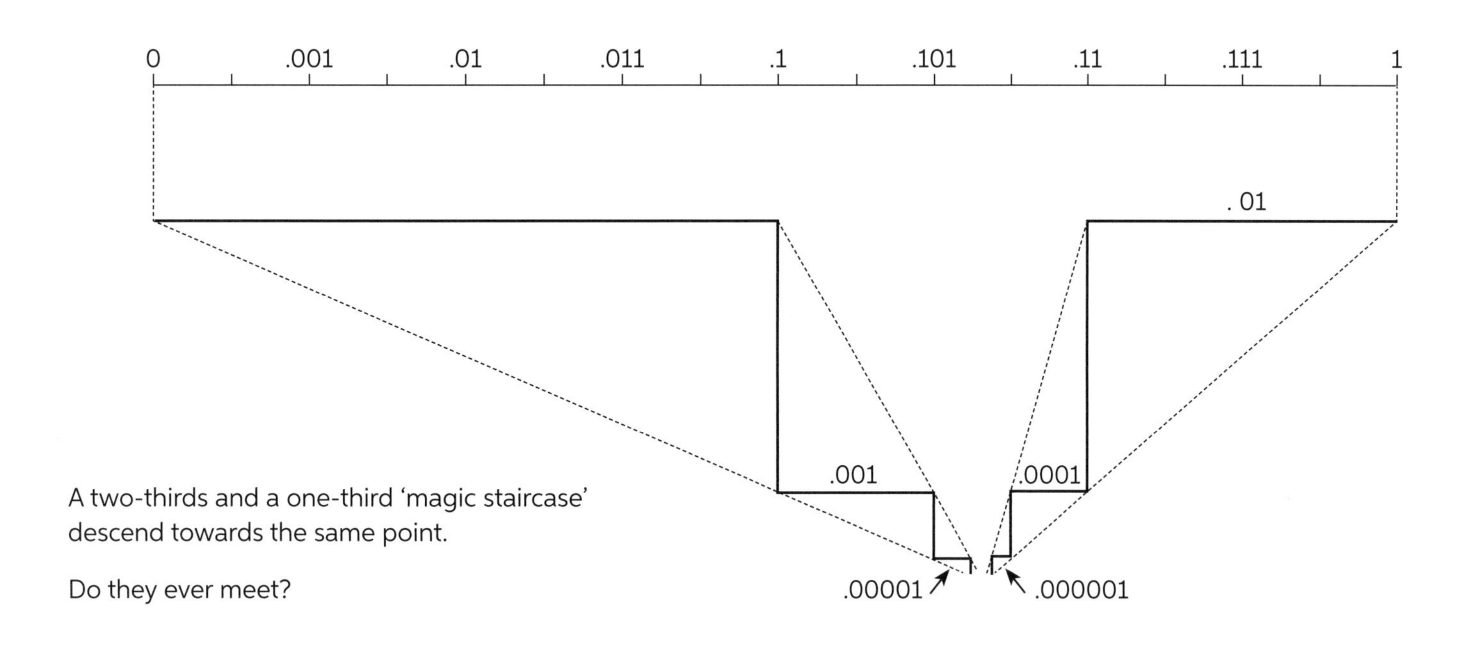


A two-thirds and a one-third ‘magic staircase’ descend towards the same point.

Do they ever meet?

## Recursive drawings

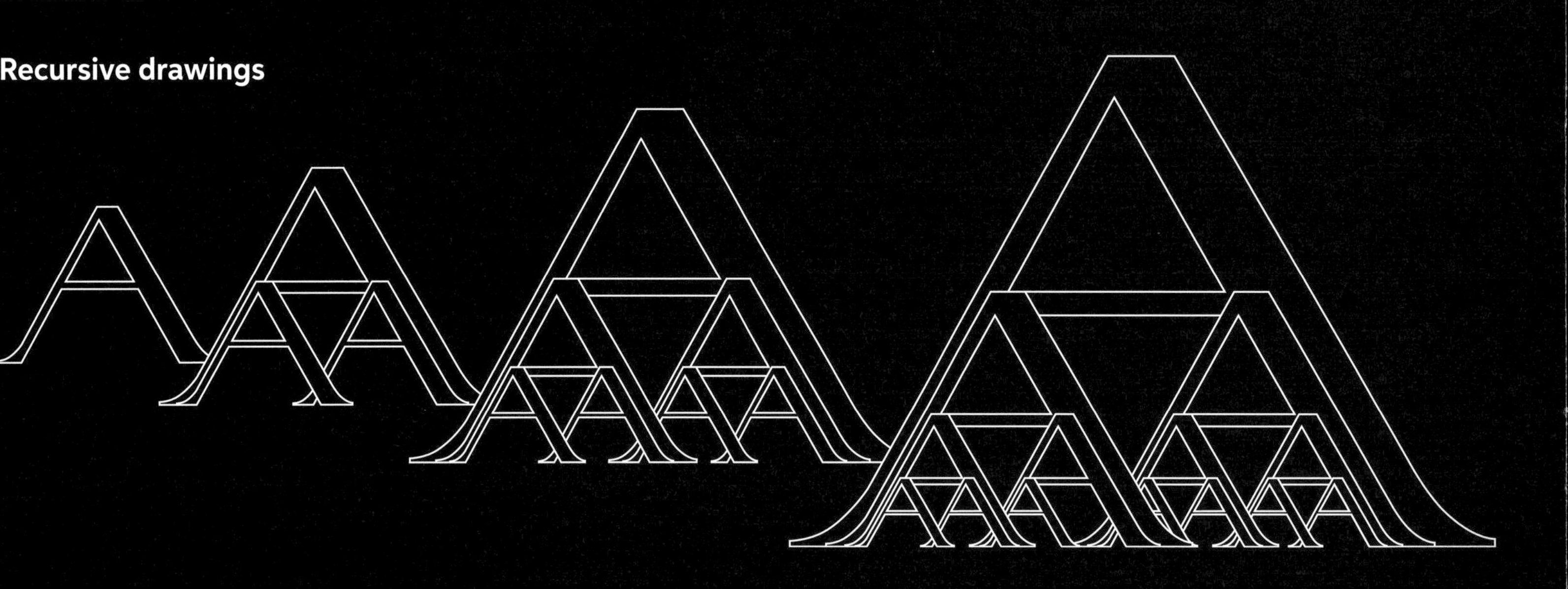

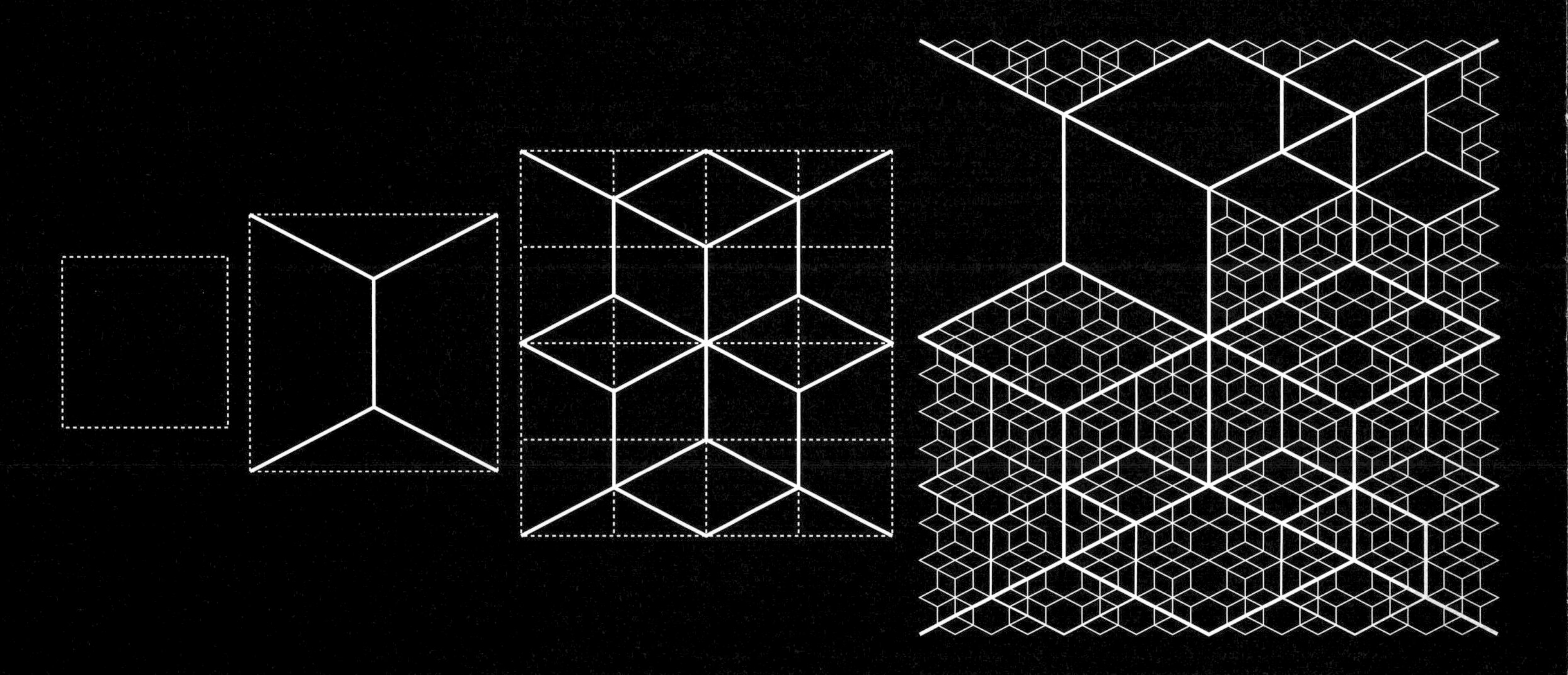

It was a dark and stormy night,
the rain came down in torrents,
there were brigands on the mountains,
and thieves
and the chief said unto Antonio:
'Tell us a story.'
And Antonio
in fear and dread of the mighty chief
began his story:
'It was a dark and stormy night,
the rain came down in torrents,
there were brigands on the mountains,
and thieves

**STAGE 1**

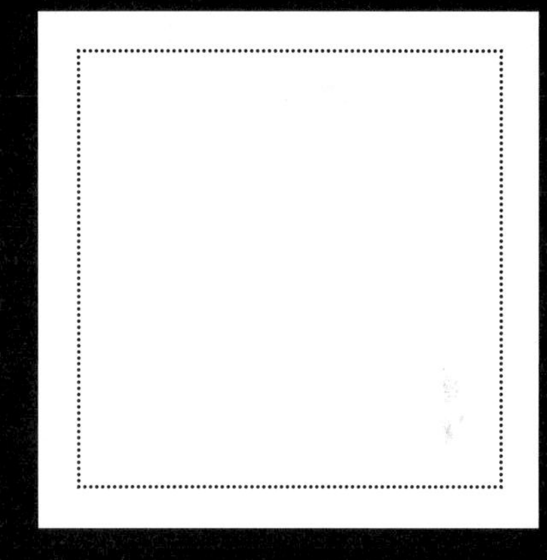

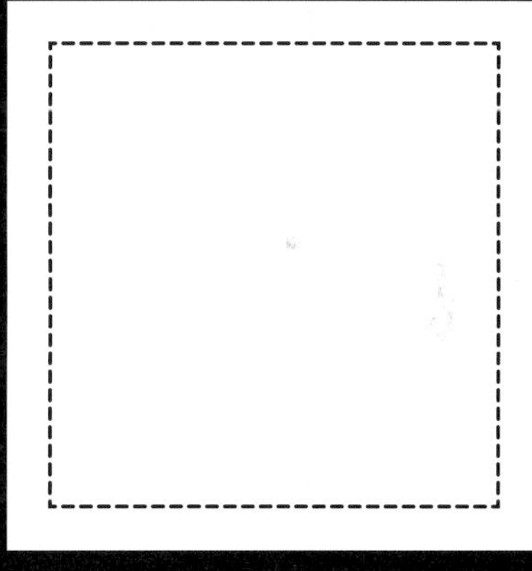

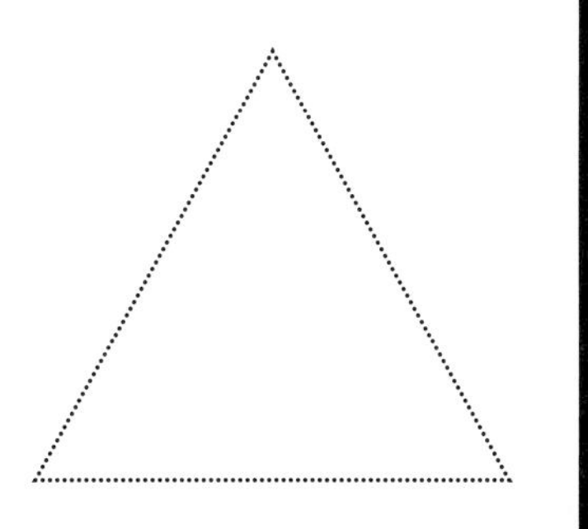

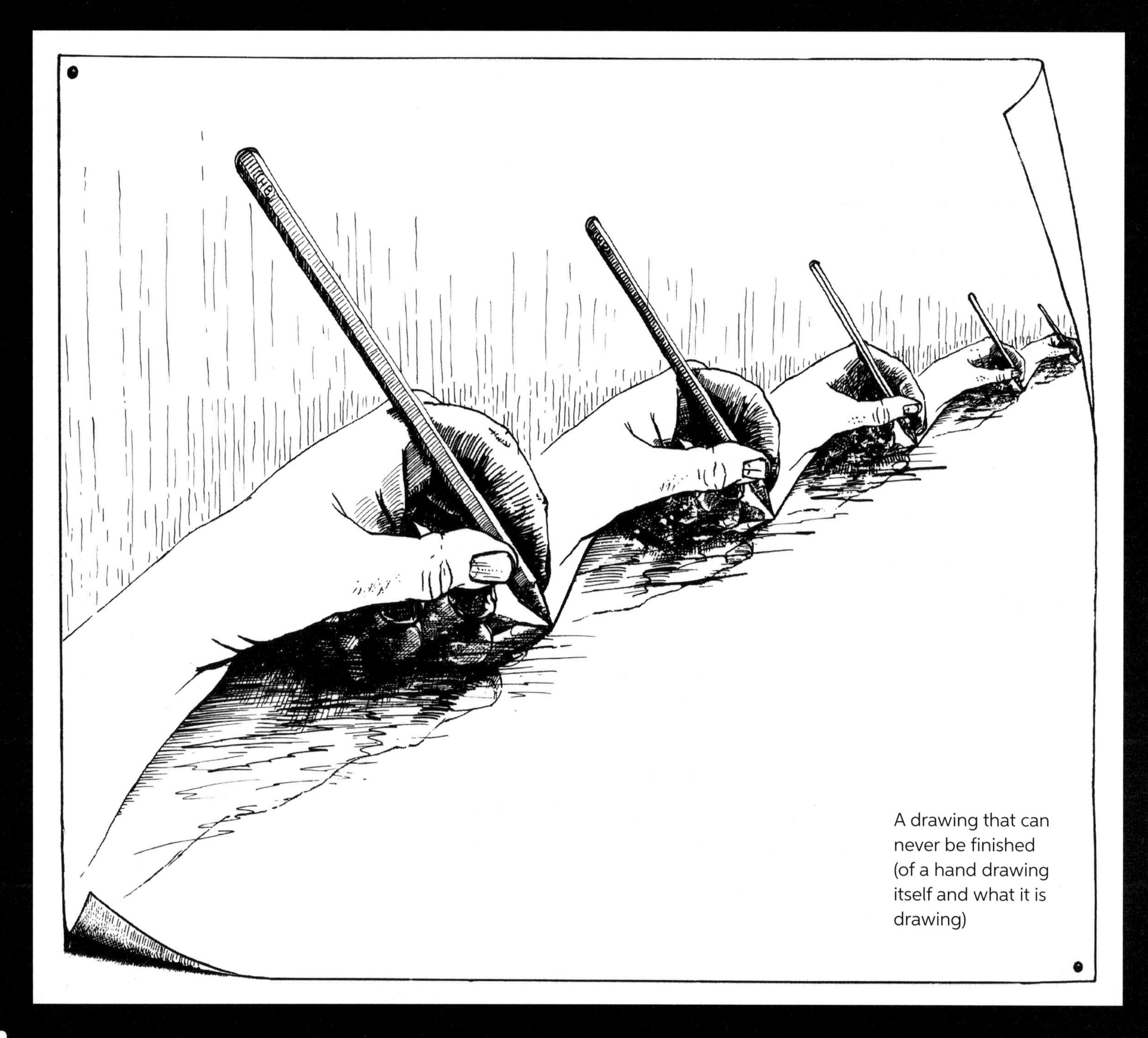


A drawing that can never be finished (of a hand drawing itself and what it is drawing)

'I sell Bibles,' he said.

Somewhat pedantically, I replied, 'In this house are several English Bibles, including the first – John Wiclifs. I also have Cipriano de Valera's, Luther's – which, from a literary viewpoint, is the worst – and a Latin copy of the Vulgate. As you see, it's not exactly Bibles I stand in need of.'

After a few moments of silence, he said, 'I don't only sell Bibles. I can show you a holy book I came across on the outskirts of Bikaner. It may interest you.'

He opened the suitcase and laid the book on a table. It was an octavo volume, bound in cloth. There was no doubt that it had passed through many hands. Examining it, I was surprised by its unusual weight. On the spine were the words 'Holy Writ' and, below them, 'Bombay'.

'Nineteenth-century, probably,' I remarked.

'I don't know,' he said. 'I've never found out.'

I opened the book at random. The script was strange to me. The pages, which were worn and typographically poor, were laid out in double columns, as in a Bible. The text was closely printed, and it was ordered in versicles. In the upper corners of the pages were Arabic numbers. I noticed that one left-hand page bore the number (let us say) 40,514 and the facing right-hand page 999. I turned the leaf; it was numbered with eight digits. It also bore a small illustration, like the kind used in dictionaries – an anchor drawn with pen and ink, as if by a schoolboy's clumsy hand.

It was at this point that the stranger said, 'Look at the illustration closely. You'll never see it again.'

I noted my place and closed the book. At once, I re-opened it. Page by page, in vain, I looked for the illustration of the anchor. 'It seems to be a version of Scriptures in some Indian language, is it not?' I said to hide my dismay.

'No,' he replied. Then, as if confiding a secret, he lowered his voice. 'I acquired the book in a town out on the plain in exchange for a handful of rupees and a Bible. Its owner did not know how to read. I suspect that he saw the Book of Books as a talisman. He was of the lowest caste; nobody but other untouchables could tread his shadow without contamination. He told me his book was called the Book of Sand, because neither the book nor the sand has any beginning or end.'

The stranger asked me to find the first page.

*continued on page 41*

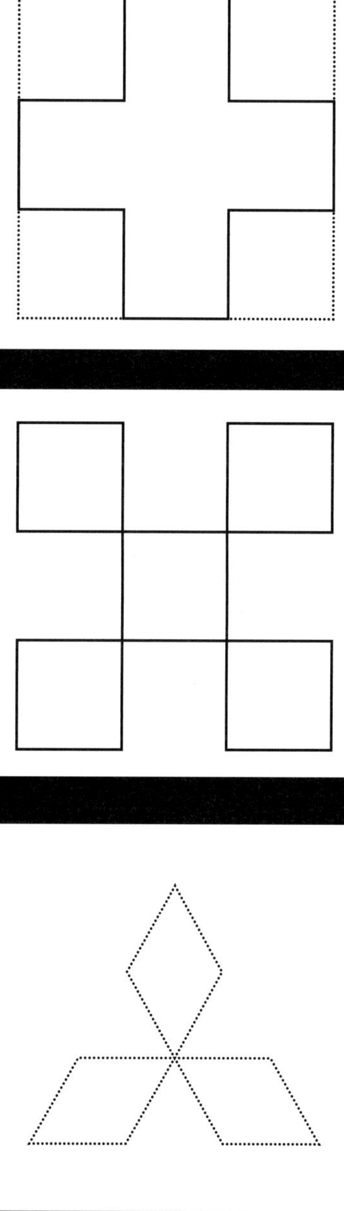

## Prime plots

Numbers are put onto a square grid in some regular way.

Prime numbers are marked.

Some patterns seem to form.

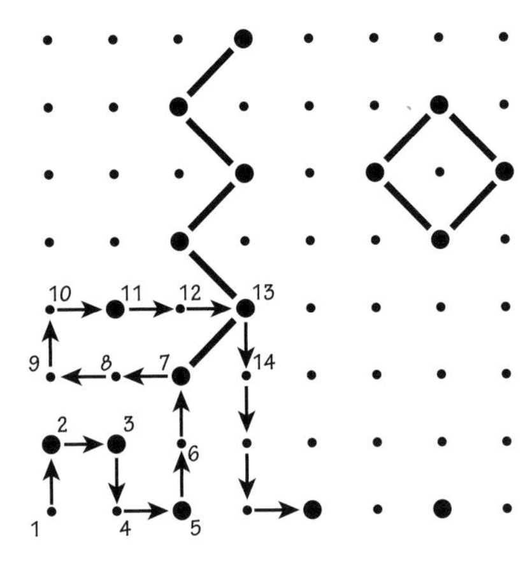


A plot like this could be started at any number.

The computer which produced the picture on the right was programmed to start its plot at 41 (bottom left).

Make a small window (about 1cm square) in a piece of card. Put this at the bottom left corner of the computer plot. Count the dots (primes) in the window. Move the window diagonally upwards. Make another count. Move it again.

Count again. (See next page.)

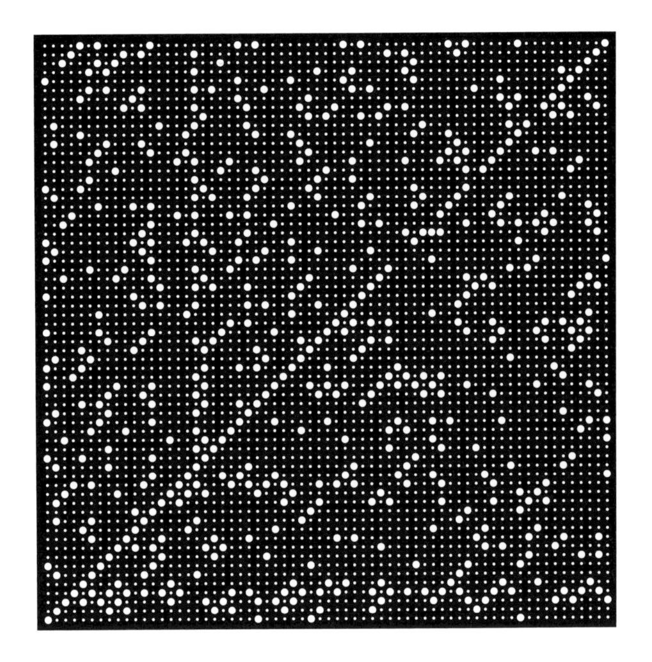

The plot of primes on the right has been made on a triangular grid.

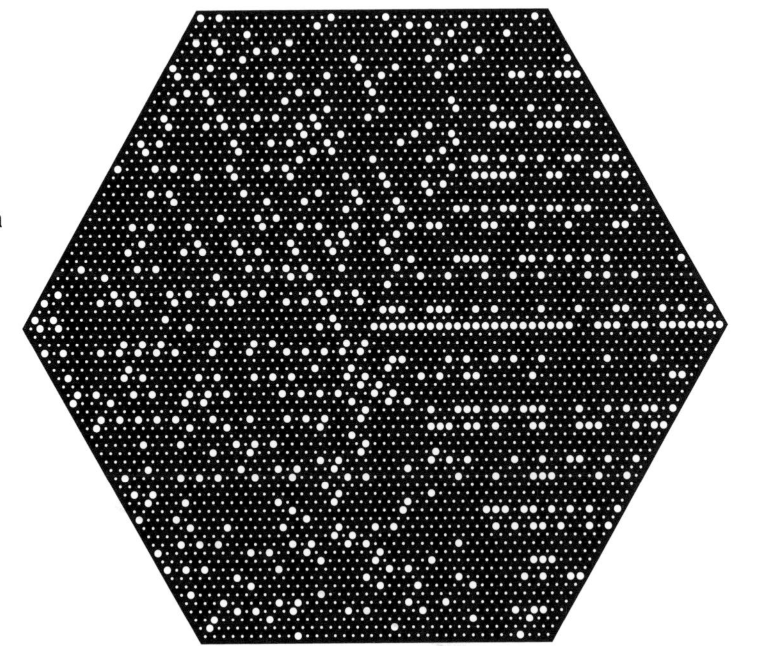

It starts in the centre at 23 and spirals out in hexagons as below:

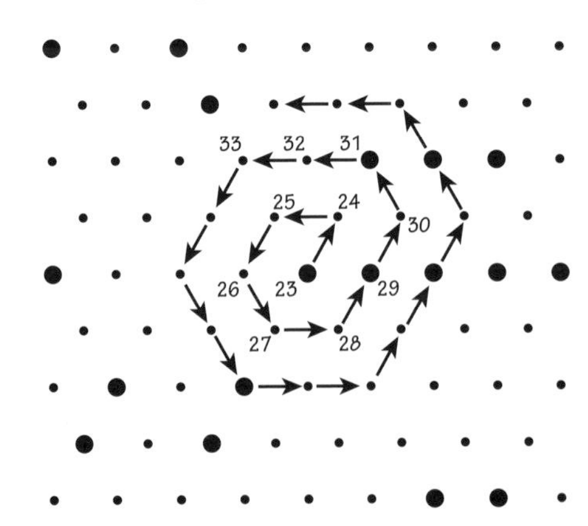

## The largest prime?

The window experiment on the page opposite suggests that as you go into large numbers … 1000, 2000, 3000, … the number of primes in each thousand gets gradually smaller.

Like the number of molecules of oxygen as you go higher and higher.

Between 1,000,000,000,000 and 1,000,000,001,000 you could expect to find very, very few primes.

Is it possible then that they would eventually peter out altogether? Is there a last prime number, beyond which there are no more to be found? It would certainly be very large. For instance, there is a number ($2^{132849} - 1$) which is known to be prime.

This number has more than a third of a million digits: so, written out in full, it would fill about 3000 books of this size.

Might that be the largest prime number? Well, no! It was shown by Euclid (more than 2000 years ago) that there is always a prime larger than any prime you like to nominate.

Euclid's argument is applied here against someone who says that 13 is the largest prime. It could be applied in the same way against an assertion that $2^{132849} - 1$ is the largest prime.

*I think 13 is the largest prime.*

**Ridiculous! You mean 2, 3, 5, 7, 11 and 13 are the only primes?**

That's right.

**What about (2 × 3 × 5 × 7 × 11 × 13) + 1?**

That's... 30031. What about it?

**Well it doesn't divide by any of your primes... 2, 3, 5, 7, 11, 13... they each leave a remainder 1.**

So 30031 is prime?

**Either that, or it has a prime factor larger than 13. So 13 is certainly not the largest prime.**

Oh well, ... perhaps 30031 is.

**Same thing... Think about (2 × 3 × 5 × ... × 30031) + 1.**

### STAGE 3

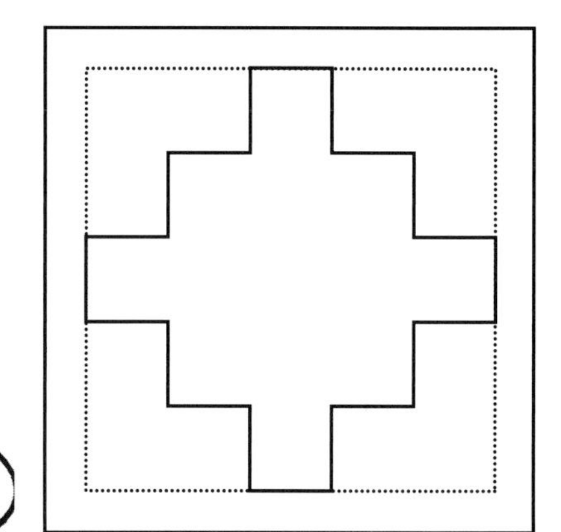

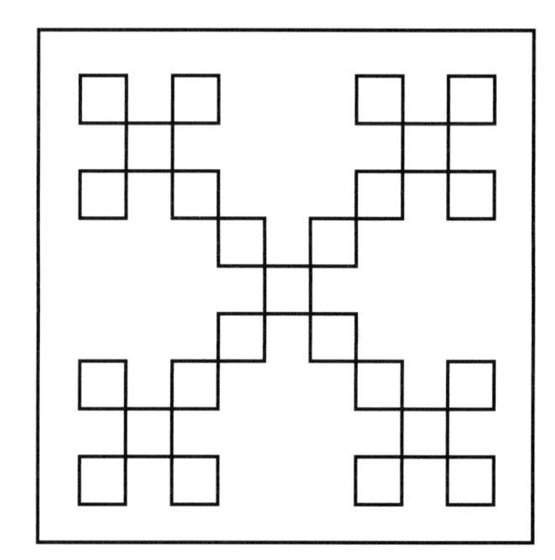

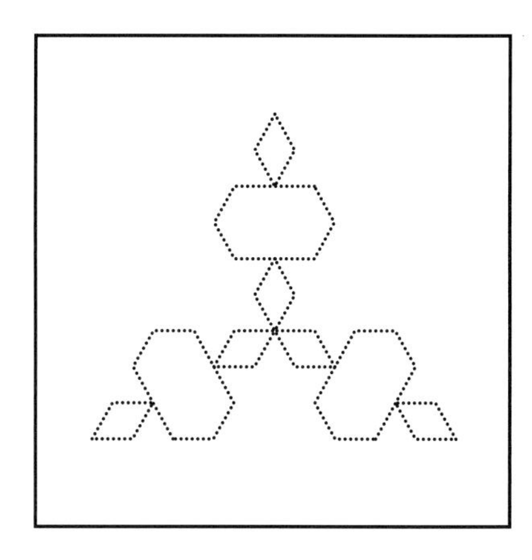

*Danilo's snail*

## Images from Italy

Young children talking ...

**Sergio** The numbers are infinite because when they reach a million they must go on or else if they stopped, what kind of world would that be?

**Simone** Infinity is like saying ... there is a corridor which never ends.

**Leo** Infinity is like a cupboard which goes even higher than a house, which never ends, never, never.

**Emi** There is a ladder which reaches to the clouds ... and the numbers can never end and if they ended at a million and someone were to go to the bank and get out some money and asked for two million and then they would have no more to pay because the numbers would have finished.

**Danilo** There is the sea and then it rains and it snows . . . and the sea never ends.

**Tizi** The stars, they are infinite.

**Leo** Infinity, that could be flowers which go on for ever, and you take the seeds of the flowers and you throw them and always more flowers come.

**Gabri** There are children on the moon who ask when the voyage is going to end. Their mother says 'Au revoir'. And the spacecraft never ends its journey.

**Cicci** The world turns round, and it turns round and it never finishes turning.

**Simone** And then you get back to the point you started at.

**Fiore** The sky never ends, the sun never ends.

**Tizi** It's not true that the earth turns for ever because if the end of the world comes the world will not turn any more and it will explode.

**Pippa** There is a thing which is truly infinite and that is the sand at the seaside because the grains of sand are so small that to fill the beach there would be not enough of them if there were millions.

**Danilo** The snail is in the middle of a labyrinth and the snail never gets out because he is small and the paths are so many and infinite ...

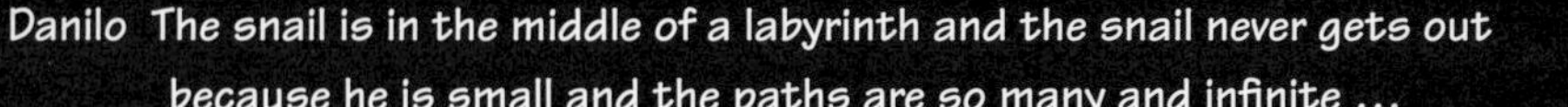

STAGE 4

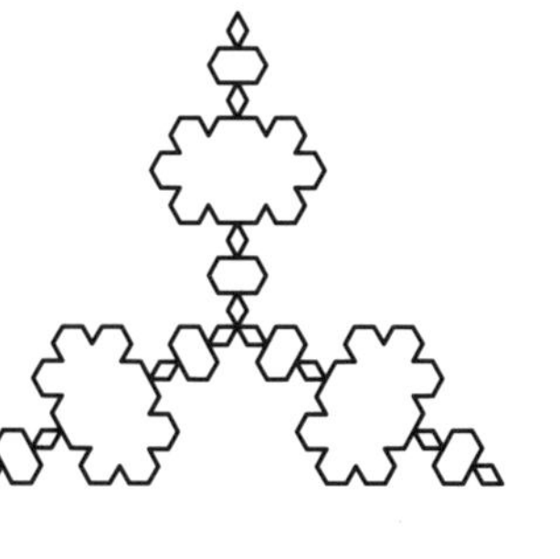

About the snowflake curve, an Italian mathematician, Ernesto Cesaro, wrote:

'What strikes me above all about the curve is that any part is similar to the whole. To try to imagine it as completely as possible, it must be realised that each small triangle in the construction contains the whole shape reduced by an appropriate factor.

And this contains a reduced version of each small triangle which in turn contains the whole shape reduced even further and so on ... to infinity—'

Cesaro went on:

'This endless embedding of the shape in itself gives us an image of what Tennyson calls the inward infinity, which is after all the only infinity we can conceive in Nature.

It is this self-similarity in all its parts, however small, that makes the curve seem so wondrous.

If it appeared in reality it would not be possible to destroy it without removing it altogether, for otherwise it would ceaselessly rise up again from the depths of its triangles like the life of the universe itself.'

**STAGE 5**

At each stage the perimeter of this ‘curve’ is the same and that of the square from which its started.

But its area reduces:

$$1 \rightarrow \frac{5}{9} \rightarrow \frac{13}{25} \rightarrow \frac{41}{81} \rightarrow ?$$

At each stage the length of this curve increases by a factor $5/3$. But its area decreases by a factor $5/9$. It seems that in the limit the curve will have infinite length but zero area.

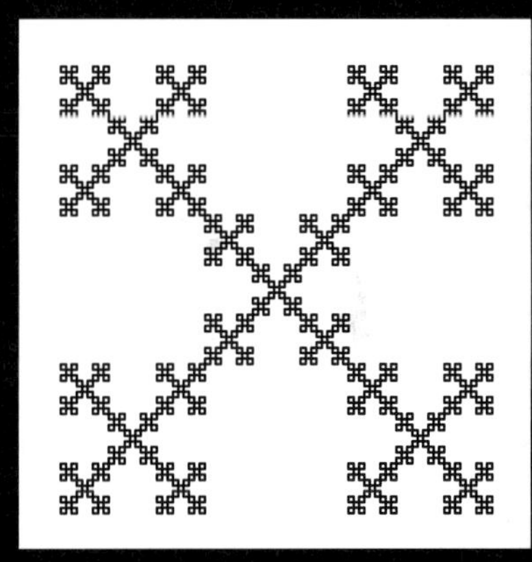

This is the ‘anti-snowflake’ curve—which tessellates with the snowflake curve.

Its perimeter increases at each stage by a factor $4/3$ and thus becomes ‘infinitely long’.

The area is always decreasing but the decrease at each stage is $4/9$ of what it was at the previous stage. This means that the area never gets less than $2/5$ of the original triangle.

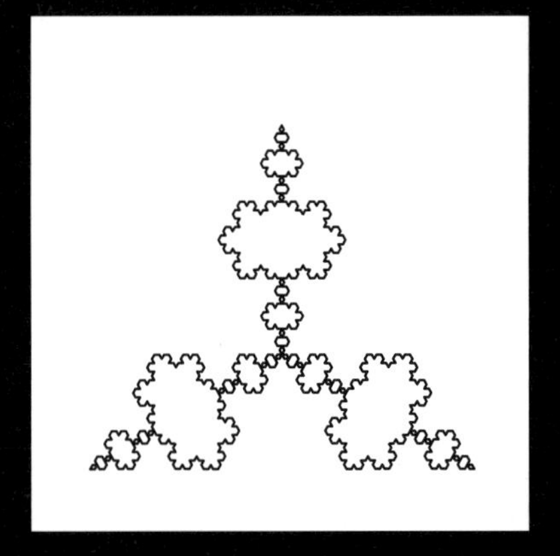

## PROGRAM FOR NESTED SQUARES

We here reproduce old code as it is easy to follow – you can easily reproduce these steps in other more modern coding languages.

```
 10  MODE1
 20  H=0:D=1000                              [sets heading & side of square
 30  PLOT4,100,0                                             [starting point
 40  PROCsquare(D)                          [calls procedure defined at 100
 50  END
100  DEF PROCsquare(D)
110  IF D<150 THEN ENDPROC                           [sets limit on recursion
120  LOCAL I
140  REPEAT                                       [repeat next 4 instructions
           PROCsquare(D*,4)                            [procedure calls itself
150                                                        with side reduced
160        PLOT 1,D*COS(H),D*SIN(H)                              [draws a line
170        H=H+RAD(90)                             [turns through right-angle
180        I=I+1
190    UNTIL I=4                              [end the "repeat" (line 130)
     ENDPROC
```

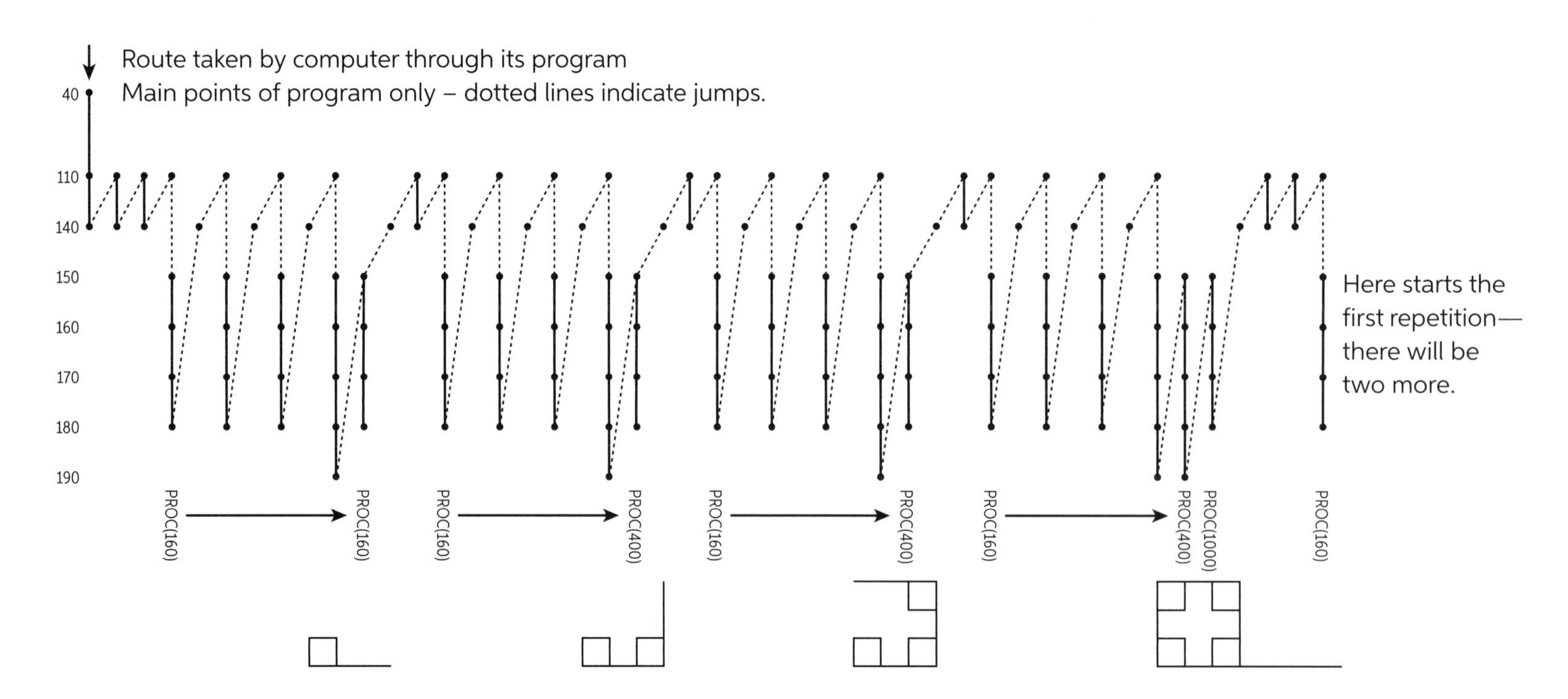

The 'PROCEDURE'
(PROCsquare) is
basically (with D = 1000)

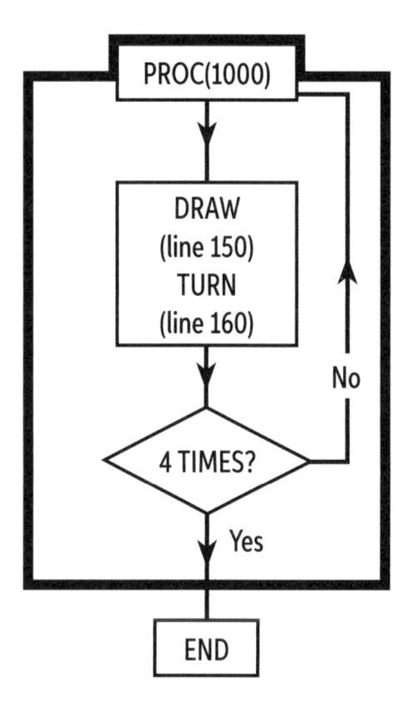


BUT before the first DRAW it meets
another PROC
with D = 400.
So now

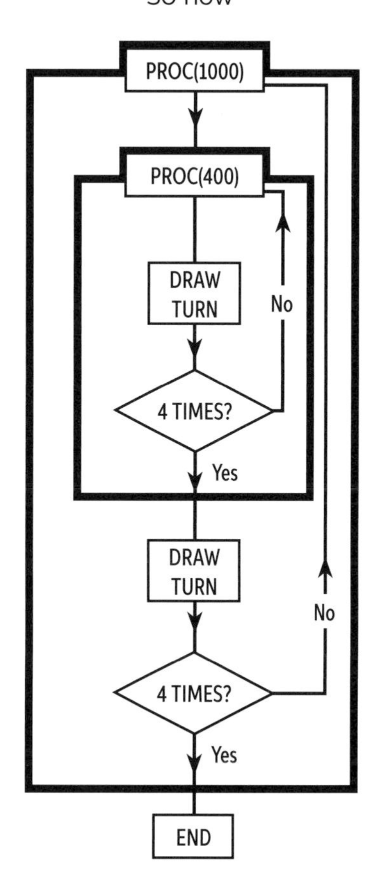


However, line 160 says that this stops
only when D<150. So PROC(400) meets
another PROC(160). Then:

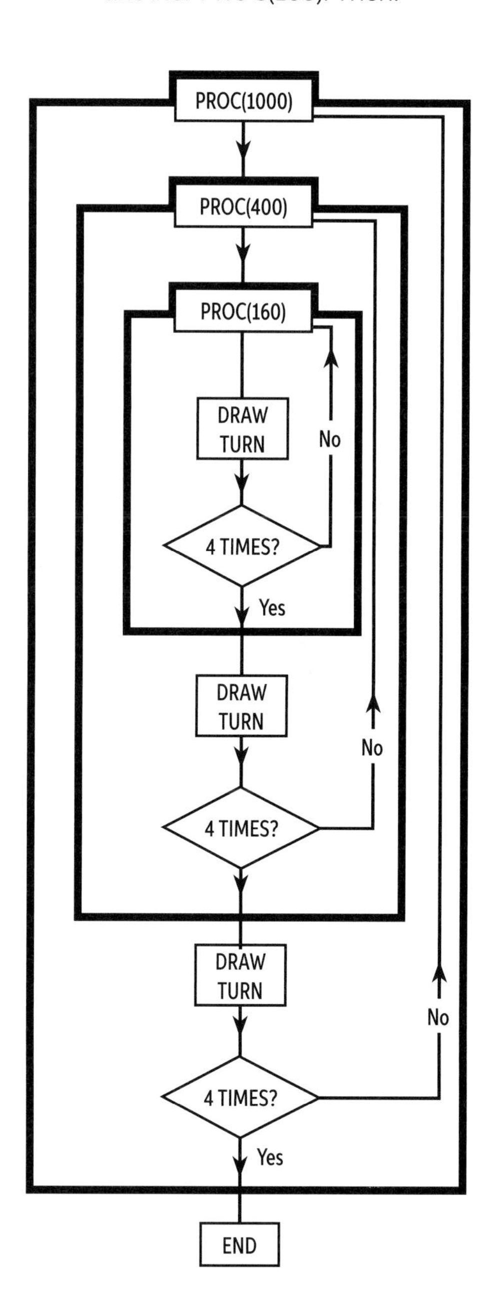

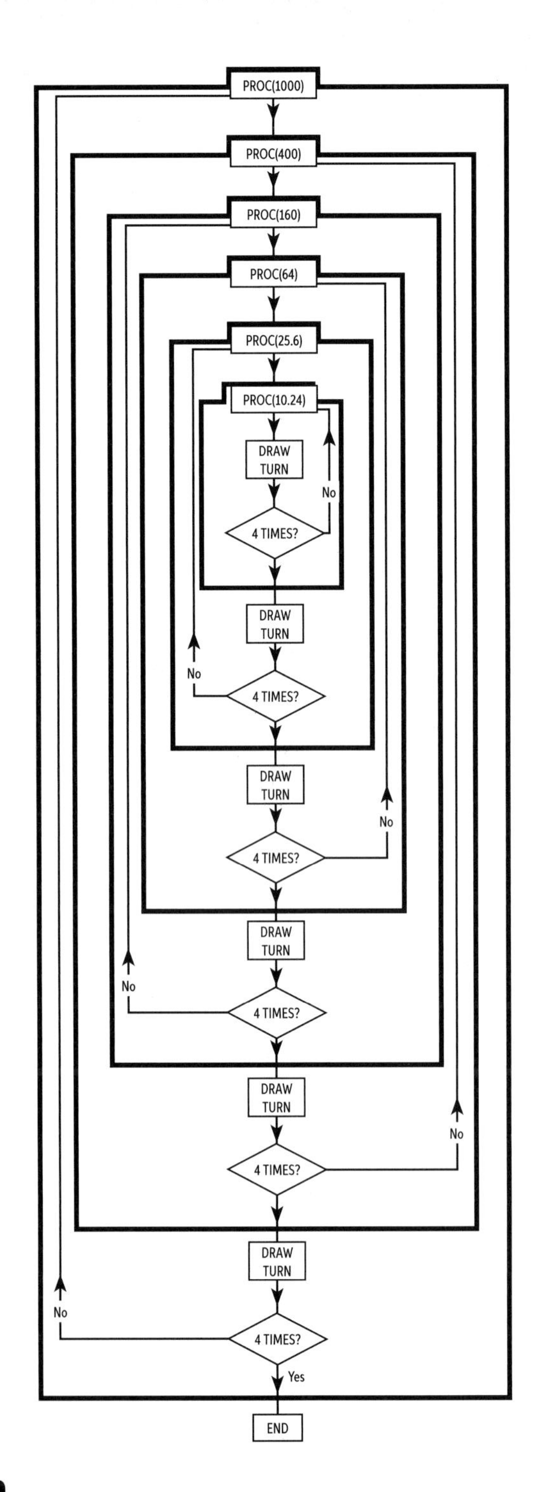


If the minimum value of D (line 110) is lowered, more PROCs' are nested inside each other.

In theory this could go on indefinitely.

The drawing on the right was made by the computer with line 110 reading:

```
110 IF D<10 THEN ENDPROC
```

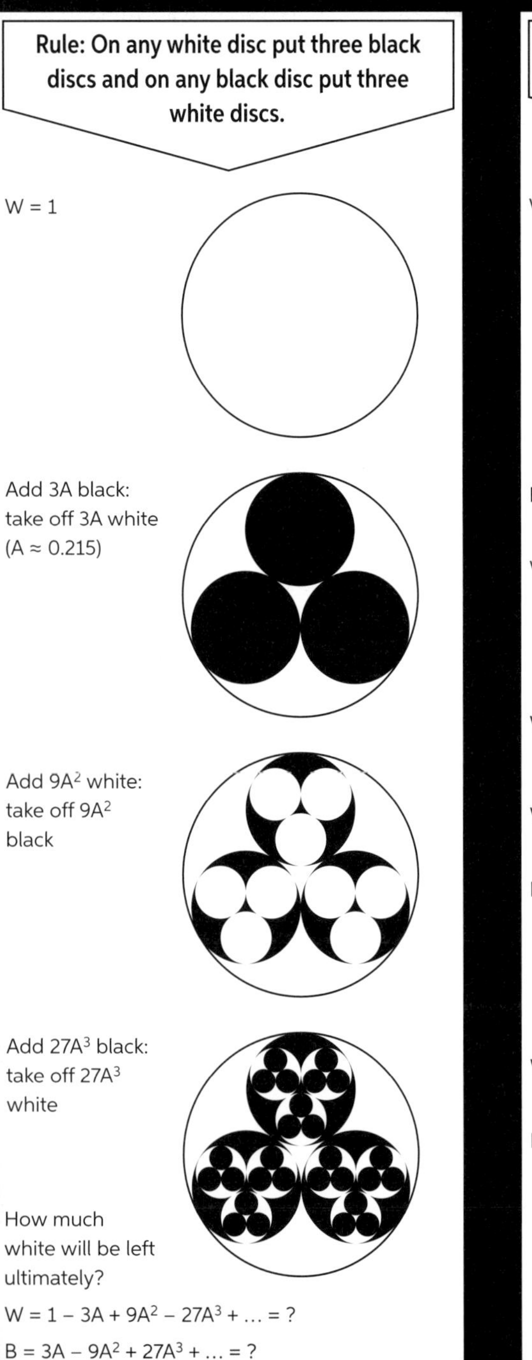
Rule: On any white disc put three black discs and on any black disc put three white discs.
W = 1
Add 3A black: take off 3A white ($A \approx 0.215$)
Add $9A^2$ white: take off $9A^2$ black
Add $27A^3$ black: take off $27A^3$ white
How much white will be left ultimately?
$W = 1 - 3A + 9A^2 - 27A^3 + \ldots = ?$
$B = 3A - 9A^2 + 27A^3 + \ldots = ?$

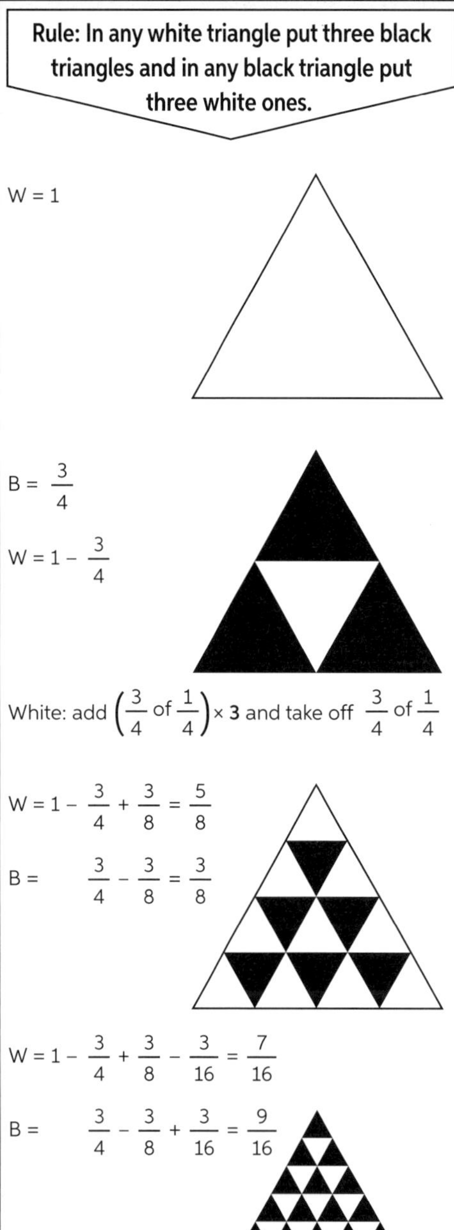
Rule: In any white triangle put three black triangles and in any black triangle put three white ones.
W = 1
$B = \frac{3}{4}$
$W = 1 - \frac{3}{4}$
White: add $\left(\frac{3}{4} \text{ of } \frac{1}{4}\right) \times \mathbf{3}$ and take off $\frac{3}{4}$ of $\frac{1}{4}$
$W = 1 - \frac{3}{4} + \frac{3}{8} = \frac{5}{8}$
$B = \frac{3}{4} - \frac{3}{8} = \frac{3}{8}$
$W = 1 - \frac{3}{4} + \frac{3}{8} - \frac{3}{16} = \frac{7}{16}$
$B = \frac{3}{4} - \frac{3}{8} + \frac{3}{16} = \frac{9}{16}$

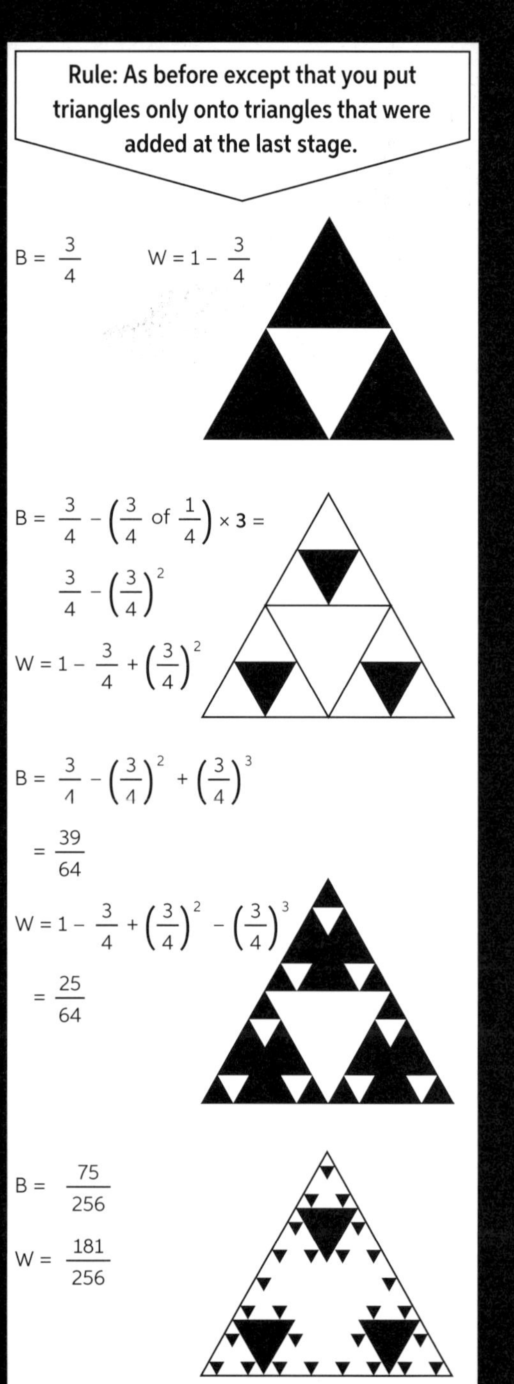
Rule: As before except that you put triangles only onto triangles that were added at the last stage.
$B = \frac{3}{4}$
$W = 1 - \frac{3}{4}$
$B = \frac{3}{4} - \left(\frac{3}{4} \text{ of } \frac{1}{4}\right) \times \mathbf{3} =$
$\frac{3}{4} - \left(\frac{3}{4}\right)^2$
$W = 1 - \frac{3}{4} + \left(\frac{3}{4}\right)^2$
$B = \frac{3}{4} - \left(\frac{3}{4}\right)^2 + \left(\frac{3}{4}\right)^3$
$= \frac{39}{64}$
$W = 1 - \frac{3}{4} + \left(\frac{3}{4}\right)^2 - \left(\frac{3}{4}\right)^3$
$= \frac{25}{64}$
$B = \frac{75}{256}$
$W = \frac{181}{256}$

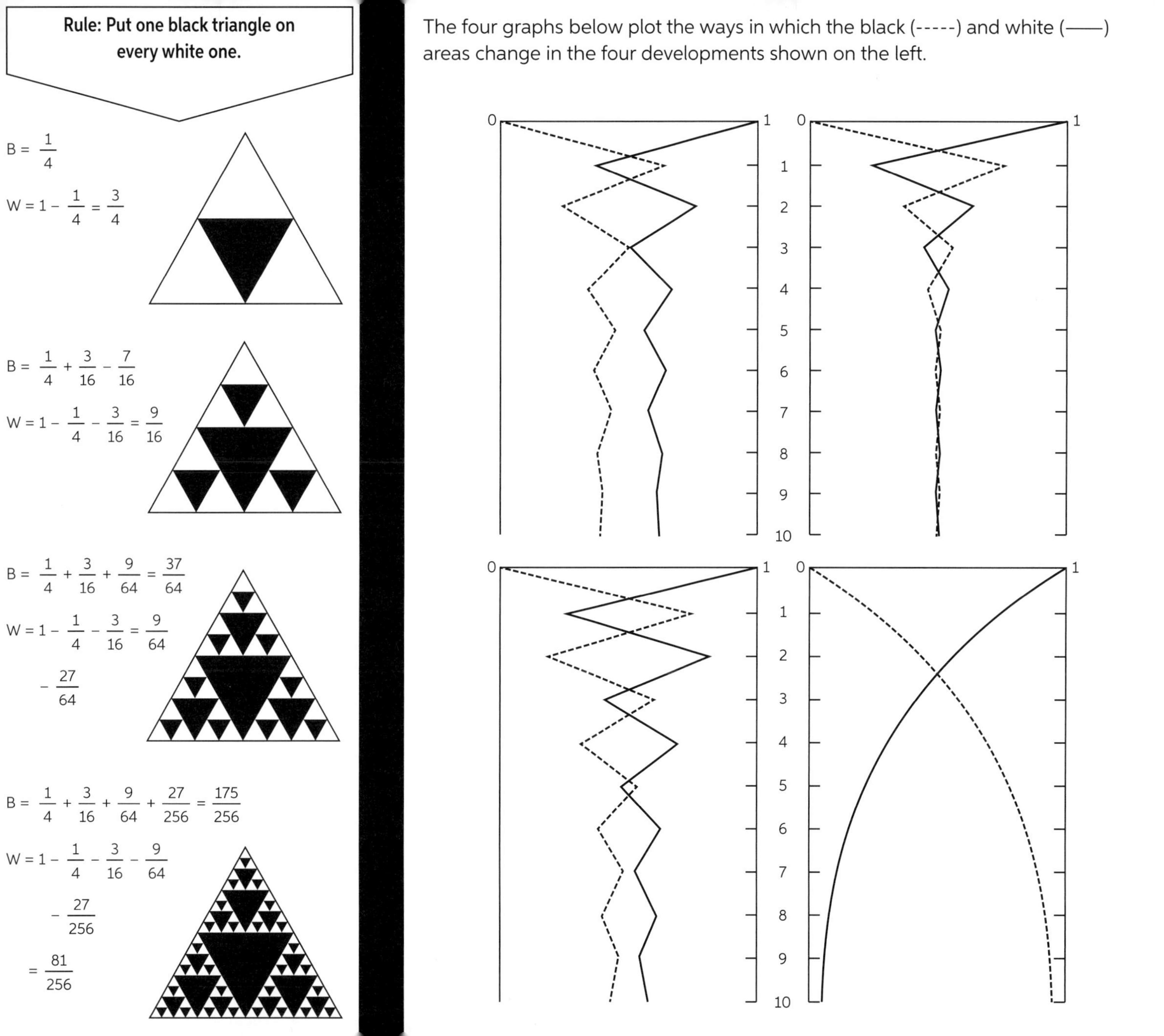

**Rule: Put one black triangle on every white one.**

$$B = \frac{1}{4}$$

$$W = 1 - \frac{1}{4} = \frac{3}{4}$$

$$B = \frac{1}{4} + \frac{3}{16} - \frac{7}{16}$$

$$W = 1 - \frac{1}{4} - \frac{3}{16} = \frac{9}{16}$$

$$B = \frac{1}{4} + \frac{3}{16} + \frac{9}{64} = \frac{37}{64}$$

$$W = 1 - \frac{1}{4} - \frac{3}{16} = \frac{9}{64} - \frac{27}{64}$$

$$B = \frac{1}{4} + \frac{3}{16} + \frac{9}{64} + \frac{27}{256} = \frac{175}{256}$$

$$W = 1 - \frac{1}{4} - \frac{3}{16} - \frac{9}{64} - \frac{27}{256} = \frac{81}{256}$$

The four graphs below plot the ways in which the black (-----) and white (——) areas change in the four developments shown on the left.

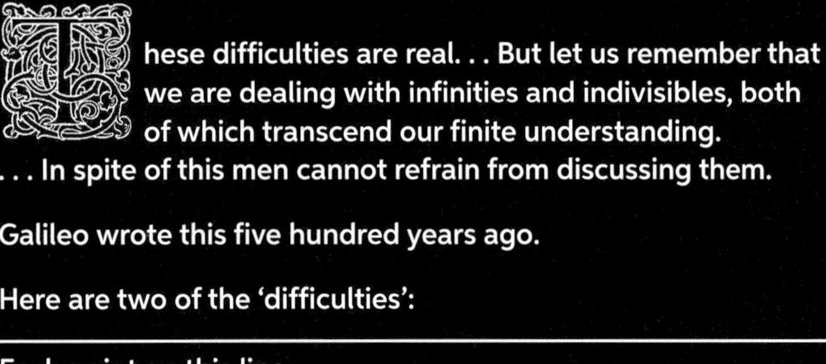

These difficulties are real. . . But let us remember that we are dealing with infinities and indivisibles, both of which transcend our finite understanding. . . . In spite of this men cannot refrain from discussing them.

Galileo wrote this five hundred years ago.

Here are two of the 'difficulties':

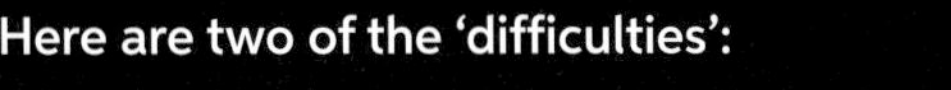

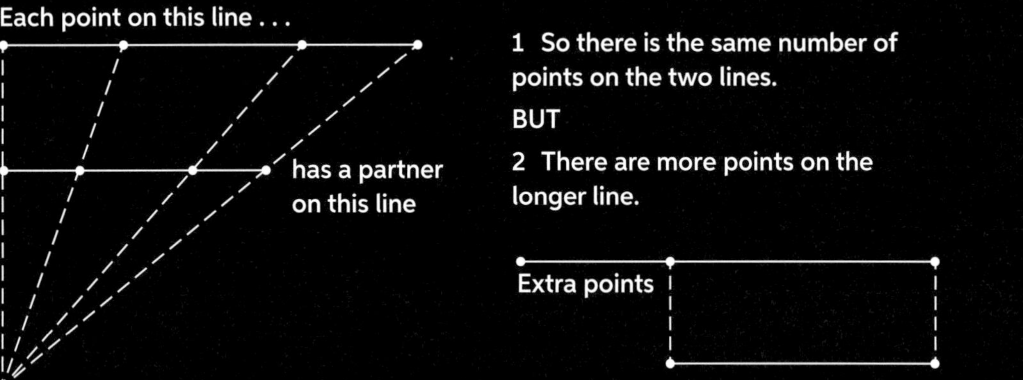


1 So there is the same number of points on the two lines.

BUT

2 There are more points on the longer line.



1 Every number has a square. So the number of squares is the *same* as the number of whole numbers.

BUT

2 All these numbers are missing from the 'squares' set. So there must be *fewer* squares than whole numbers.

{1, 2, 3, 4 …} {101, 102, 103, 104 …}

{1, 2, 3, 4 …} {10, 20, 30, 40 …}

{1, 2, 3, 4 …} {1, 10, 11, 100, 101 …}

{1, 2, 3, 4 …} {2, 3, 5, 7, 11 …}

{1, 2, 3, 4 …} {… −2, −1, 0, +1, +2 …}

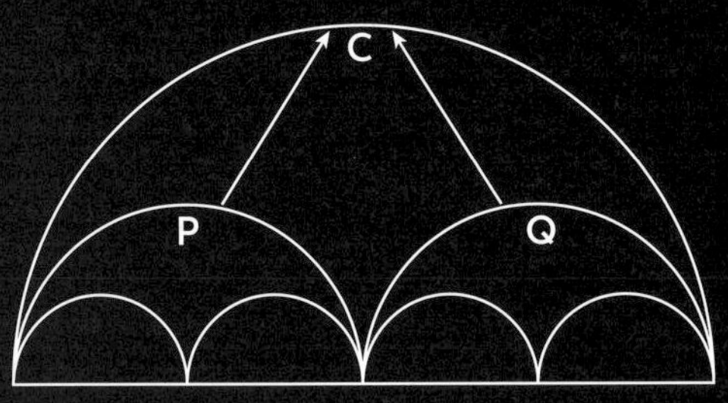


$C = \pi$ x (radius);
$P = Q = \pi \times$ (half-rad.)
So $P + Q = C$

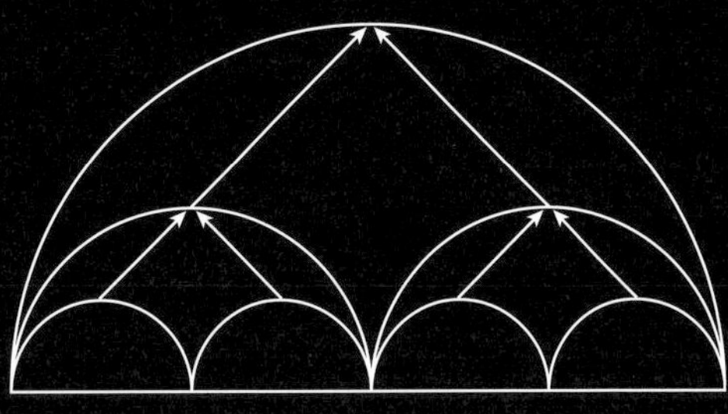

We can do this again
4 semicircles = C

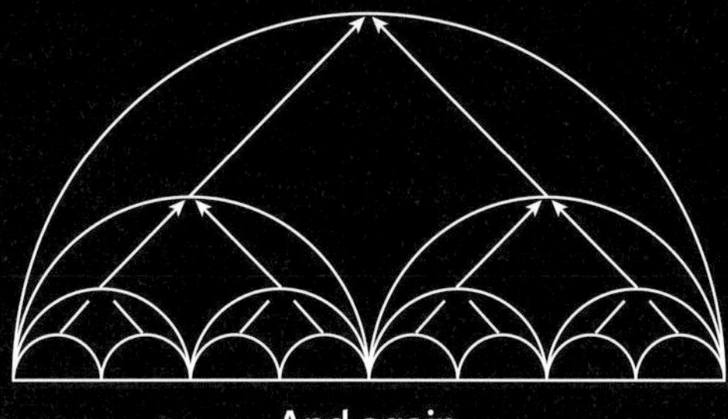

And again

and again, and again . . .

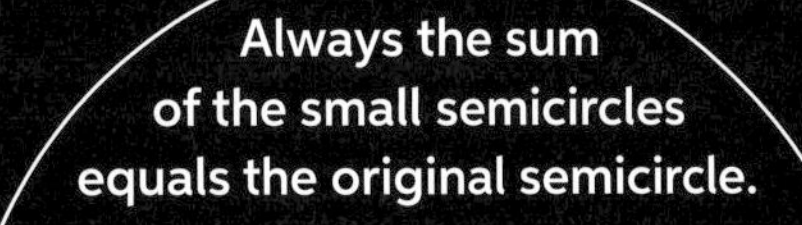

Always the sum
of the small semicircles
equals the original semicircle.

The more we do this, the closer the semicircles get to the diameter until

eventually . . .

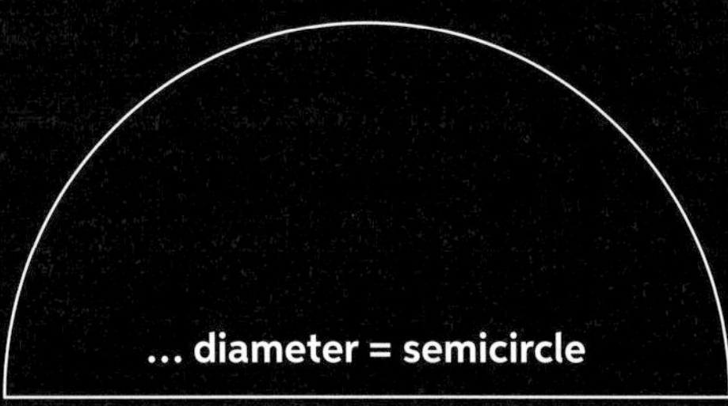


ie $2 \times$ (radius) $= \pi \times$ (radius)
So $\pi = 2$ (!)

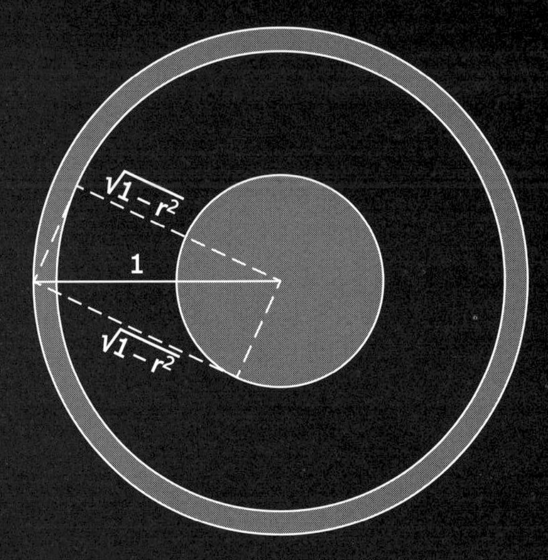


Concentric circles of radius 1, r and $\sqrt{1-r^2}$ have areas $\pi$, $\pi r^2$ and $\pi(1-r^2)$. So the shaded areas (above) are equal.

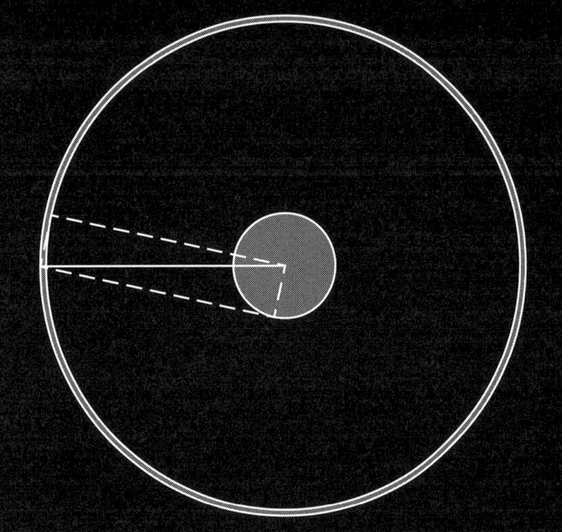

Now make r smaller and smaller . . . until . . .

... the point equals the circumference.

In Galileo's words: 'Shall we not call them equal seeing that they are the last traces and remnants of equal magnitudes?'

Being a giant, Gog liked big things. One day he brought home a huge board with slats of wood that fitted onto it. 'It's a fraction board,' he told his wife proudly (she was called Magog).

'These bits of wood show all the fractions. There's two 1/2's, three 1/3's, four 1/4's . . . humans have them in their schools but they have silly little ones — theirs stop at 1/12's. This one is an infinite one: there's no end to it.'

'Well what d'you want it for?' Magog asked him.

'Ha! I'm going to set you a problem!'

Gog took out most of the slats, throwing them into a huge pile in the garden. On the board he left just one slat of each kind.

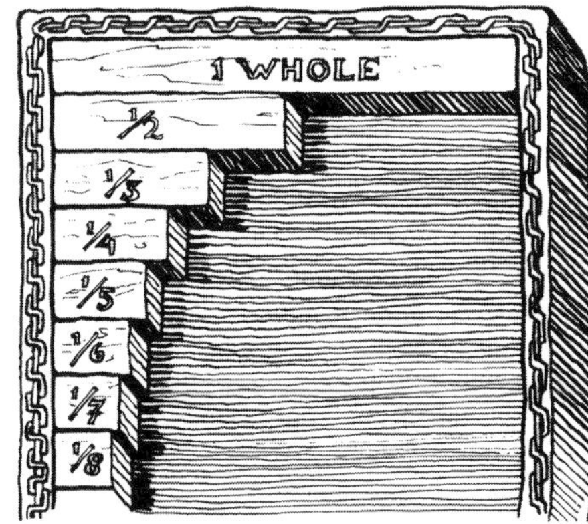


'Now what you've got to do, Magog, is to rearrange these pieces I've left so they cover the whole board. They don't have to fit exactly, but none of the board must be showing when you've finished.'

And he sat down, puffing huge clouds of smoke from his pipe. And smiling.

'Not that I expect you'll ever finish,' he added with a chuckle.

Magog thought this was a silly sort of a problem, but she was used to Gog's funny ideas and thought she'd humour him. So she started. She filled in the rest of the second row with the 1/3 piece and the 1/4 piece, and then found that the next eight pieces covered the third row. She began to feel more hopeful. Maybe she could go on like this for ever — but the pieces were getting very small.

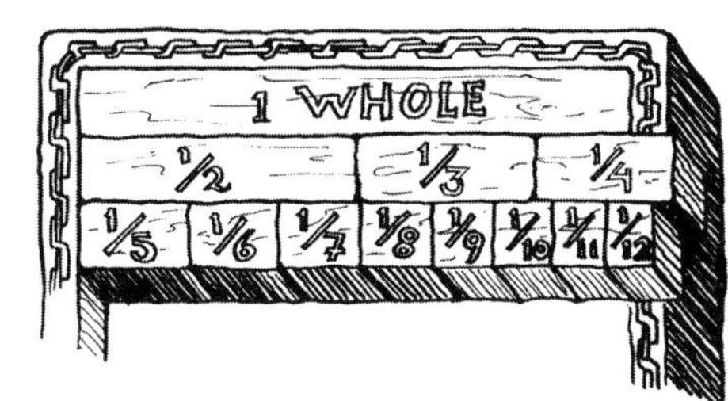


The next four went only about a quarter way across.

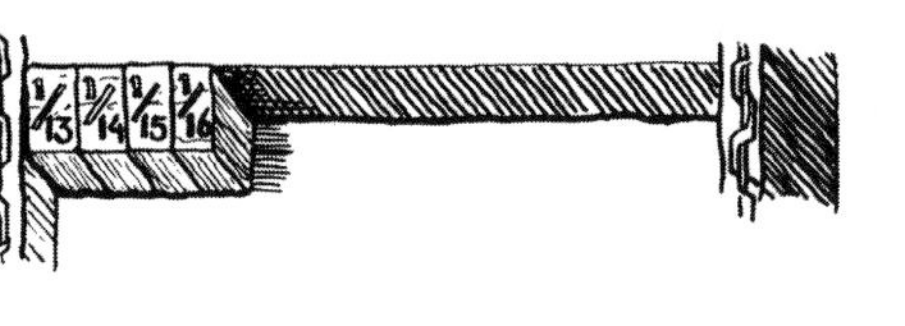


The next four would go less than a quarter way, because, she thought, each of them is less than 1/16 and 4 × 1/16 = 1/4.

But then, she thought, they are each more than 1/32 — in fact, there are 16 pieces like this (more than 1/32) and 16 × 1/32 = 1/2. That means those 16 pieces will cover half a row!

So Magog got her great idea.

'Ah-ha!' she said.

Gog stood up.

'You haven't done?' he cried.

'No,' said Magog. 'But I think I know how to do it!'

'Look at that last one I've put there . . . 1/64. Between that and the 1/32 (she pointed halfway along the last row) there are 32 pieces — 1/33, 1/34, 1/35 . . . all the way to 1/64.'

'Well?' said Gog. 'What about it?'

'You see, if they were all 1/64 they'd fill up that half-row because 32 × 1/64 = 1/2. But actually they're each bigger than 1/64. So they'll more than fill the half-row.'

'Mm . . .'

'And the same works for all the others,' Magog went on. 'I could fill the next half-row with the pieces from 1/65 to 1/128 . . . there's 64 of them each bigger than 1/128. So together they will be bigger than 1/2. So I'll always be able to fill up the next half-row because there's no end to the pieces.'

Gog had to admit that this meant she would be able to fill the whole board. Obviously he'd made the problem far too easy.

'See if you can do this one then,' he said. 'I'll take some more of them out. First, I'll take out the whole one and then I'll take out every other one — the 1/3, and the 1/5, and the 1/7, and . . .'

He hurled them onto the pile in the garden, leaving the board like this:

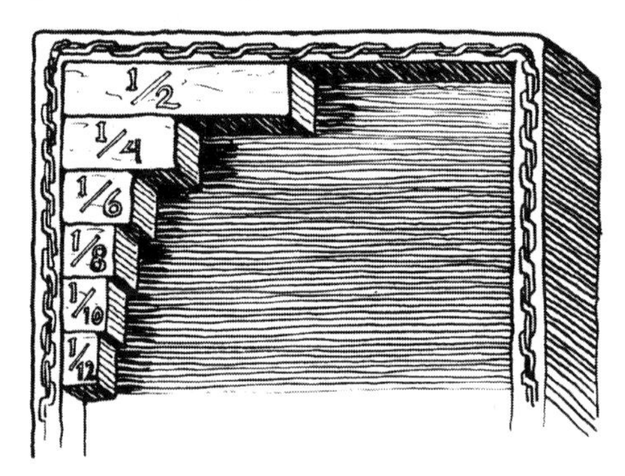


'Now see if you can fill the board with those pieces.'

Magog looked at them for a bit. And then she looked a bit longer. Suddenly it struck her that what Gog had left was

exactly half of what he had left last time. In each row there was a piece just half the size of the piece he'd left there before. 'Aha!' she thought.

'That's all right. I could cover the board with those pieces. I would need twice as many. But that doesn't matter because you've got no end of them. It'll go like this:'

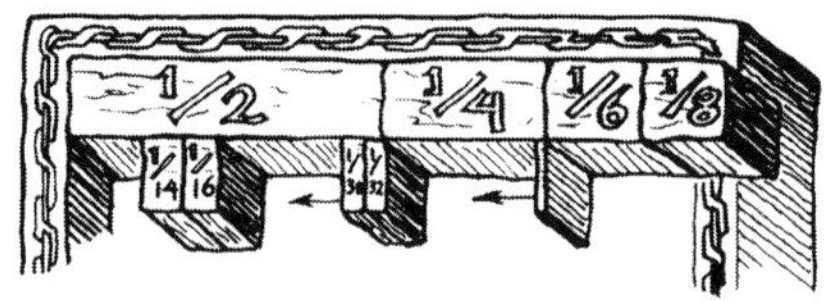


Gog was really nettled this time.

'All right,' he said. 'I'll take some more pieces out. See if you can do it with this lot.'

Again he threw out every other piece. So now the board was like this:

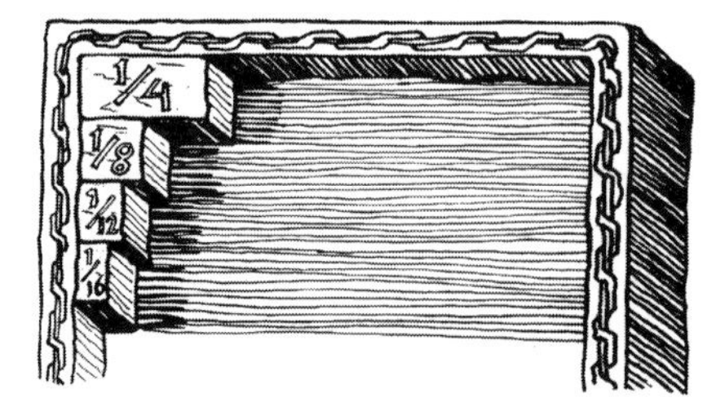


But Magog was getting rather tired of this game.

'Well actually,' she said, 'there's some wood in the garden I want to chop up for the fire. But I tell you what — I'll leave a problem for you to try.'

So before she went she left the board like this:

It certainly looked easier than the last problem he had given Magog, and Gog started happily to fill up the other half of the second row.

This row seemed to be filling up quite quickly at first, but after a while Gog realised that the next piece only filled just half of the bit left. So there was always half of that bit left uncovered.

*Then he remembered he had to take the dog for a walk.*

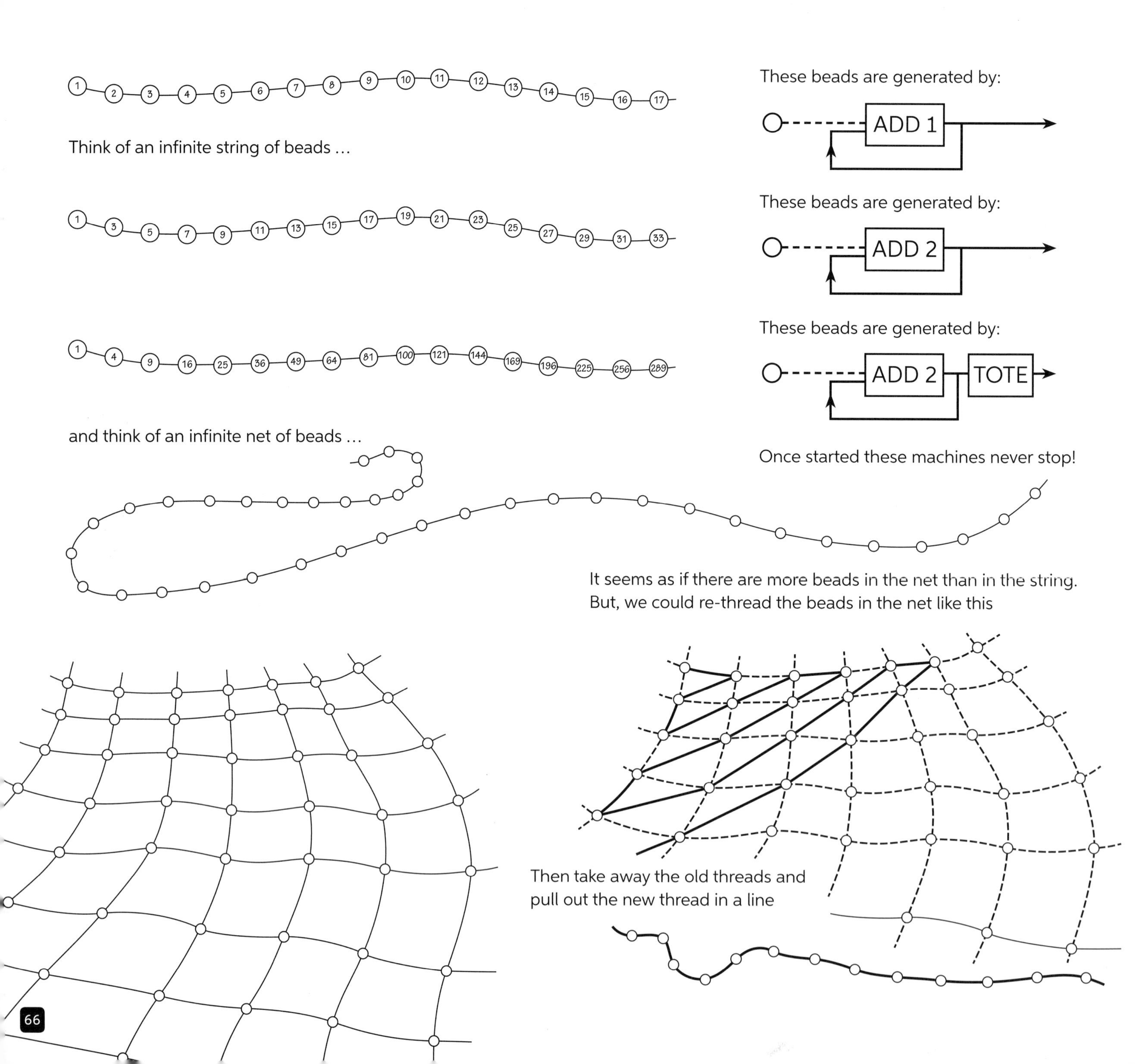
1 2 3 4 5 6 7 8 9 10 11 12 13 14 15 16 17
Think of an infinite string of beads …
These beads are generated by:
ADD 1
1 3 5 7 9 11 13 15 17 19 21 23 25 27 29 31 33
These beads are generated by:
ADD 2
1 4 9 16 25 36 49 64 81 100 121 144 169 196 225 256 289
These beads are generated by:
ADD 2
TOTE
and think of an infinite net of beads …
Once started these machines never stop!
It seems as if there are more beads in the net than in the string.
But, we could re-thread the beads in the net like this
Then take away the old threads and
pull out the new thread in a line

Is there a shortest finite string?

Is there a longest finite string?

Is there a shortest finite string?
Are all infinite strings the same length?

Is this string twice as long as the two above?

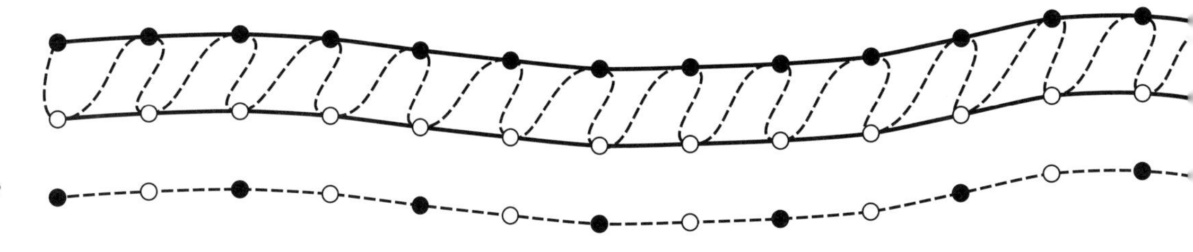

In the black-and-white string,

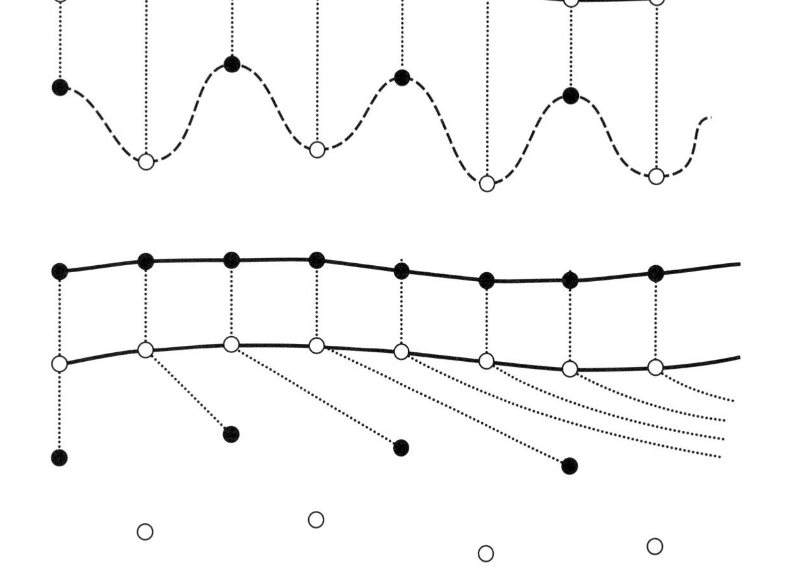

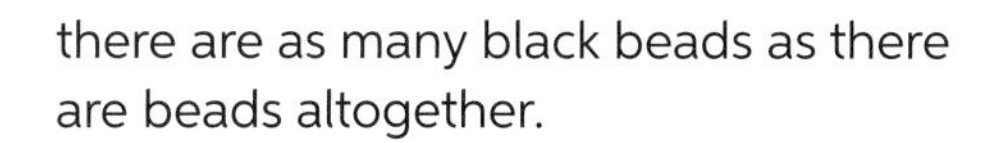

there are as many black beads as there are beads altogether.

(There are as many even numbers as there are numbers altogether. Does this mean that all numbers are even?)

Are these two infinite strings the same length?

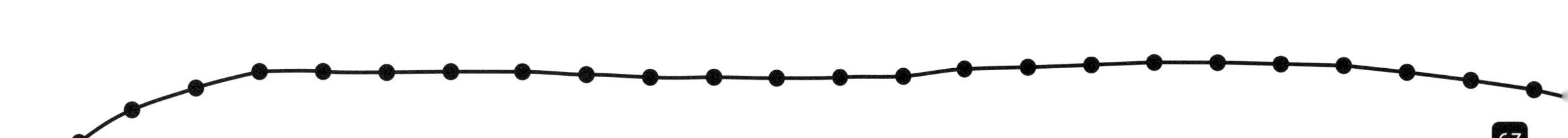

## Presque tout

How many numbers have at least one digit a 9?

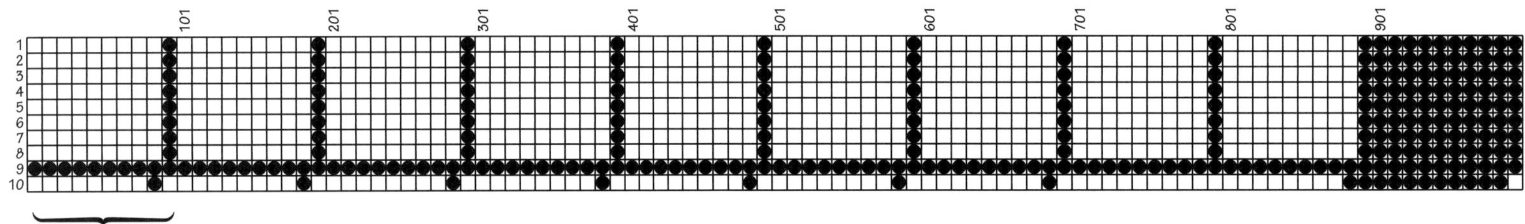


An argument (which shows that almost all numbers have a 9-digit!):

If these numerals were numbers in base-9, they would *look* just the same except that all those with a 9 in them would be omitted.

In base 10: 1 2 3 4 5 6 7 8 9 10 … 17 18 19 20 21 … 87 88 89 90 91 92 93 94 95 96 97 98 99 100

In base 9: 1 2 3 4 5 6 7 8 ↓ 10 … 17 18 19 20 21 … 87 88 ↓ ↓ ↓ ↓ ↓ ↓ ↓ ↓ ↓ ↓ ↓ 100

These have no partners … … because they contain a 9-digit.

The others are paired off if they look the same: they may not *mean* the same.

In fact, 100 in base-9 means $9^2 = 81$.

So there are 81 numbers in the base-9 line, as against 100 in the base-10 line.

- The difference, $100 - 81 = 19$, is the number of numbers up to 100 which have a 9 in them.

Now imagine the same thing but with the base-10 line ending at 1000 (ie $10^3$). The base-9 line would also end at a number written 1000, but meaning $9^3$ (= 729).

- Again, the difference, $10^3 - 9^3 = 271$, is the number of numbers up to 1000 which have a 9 in them.

We could go on like this:
up to 10,000, there are $10^4 - 9^4$ numbers with a 9 in them.
up to 1000,000 there are $10^5 - 9^5$ numbers with a 9 in them.

And so on.

- Put it like this:
  of the first 100 numbers, 19 have a 9 in them (ie about 1/5, of them)
  of the first 1000 numbers, 271 have a 9 in them (ie over 1/4 of them)
  of the first 10000 numbers, 3439 have a 9 in them (ie about 1/3 of them)
  You see that the proportion of numbers with a 9 in them is growing.

It goes on growing:

| For numbers up to: | 10 | $10^2$ | $10^3$ | $10^4$ | $10^5$ | $10^6$ | $10^7$ | $10^8$ | $10^9$ | $10^{10}$ | | $10^{15}$ | | $10^{20}$ | | $10^{32}$ | | $10^{64}$ |
|---|---|---|---|---|---|---|---|---|---|---|---|---|---|---|---|---|---|---|
| Proportion with 9: | .1 | .19 | .27 | .34 | .41 | .47 | .53 | .57 | .61 | .65 | | .79 | | .88 | | .97 | | .99 |

- Is it a fair conclusion that, considering ALL the numbers, *almost all of them have a 9 in them*?

The conclusion 'almost all numbers contain a 9' is difficult to believe.
Do these diagrams make it more believable?

In the first 100 numbers there are 19 with a 9.

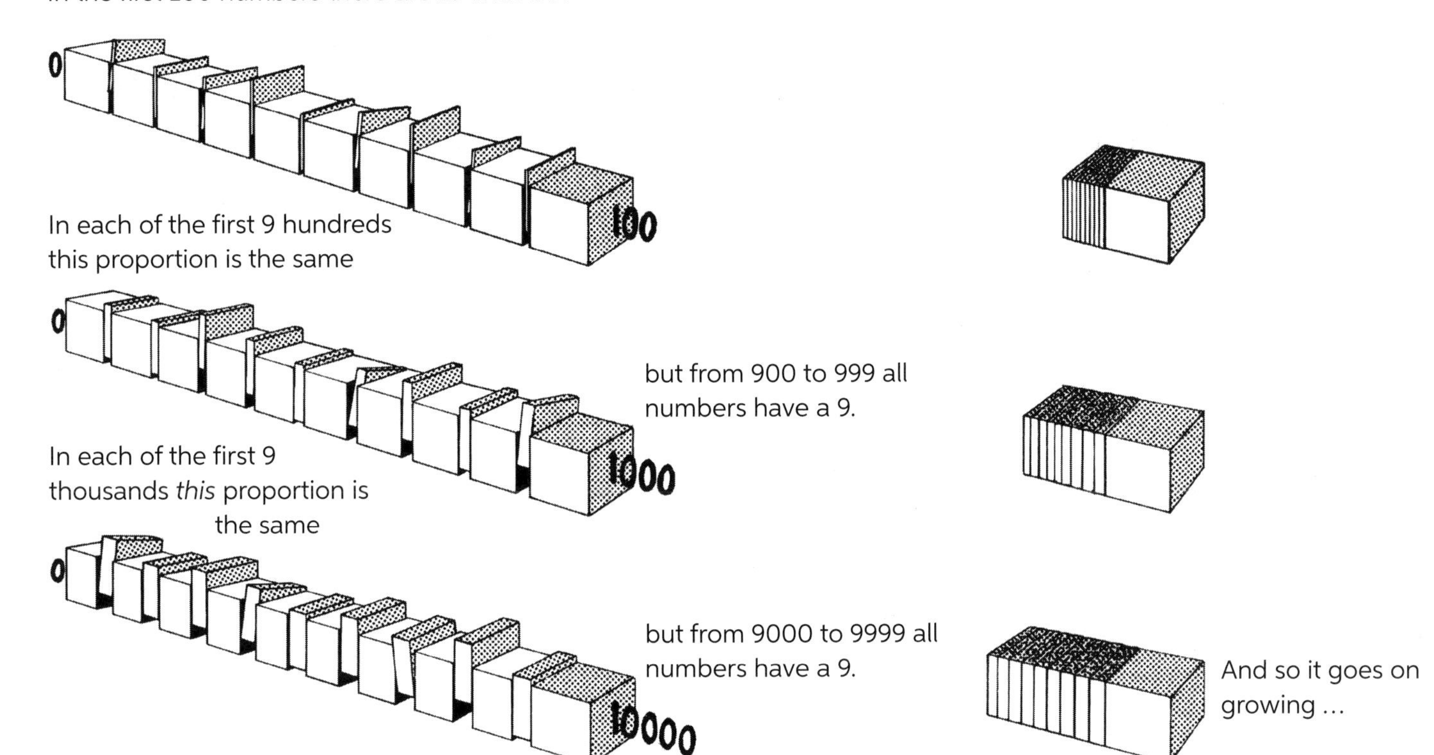

0
100
In each of the first 9 hundreds
this proportion is the same
0
1000
but from 900 to 999 all
numbers have a 9.
In each of the first 9
thousands *this* proportion is
the same
0
10000
but from 9000 to 9999 all
numbers have a 9.
And so it goes on
growing …


- Is there anything so very special about a 9 (as compared with an 8, say)?

  To every number with a 9-digit, there corresponds a number with an 8-digit:

  29 ⇄ 28
  9 ⇄ 8
  197 ⇄ 187
  3985 ⇄ 3895 etc

- So, if 'almost all numbers contain a 9', it must also be true that 'almost all numbers contain an 8'. And similarly almost all numbers contain a 7 (or a 6, or a 5 …)

- Concisely: In the first $10^N$ numbers, the proportion which contain a chosen digit (1, 2, 3, … 8 or 9) is $1 - \left(\frac{9}{10}\right)^N$.

  For the very large N, $\left(\frac{9}{10}\right)^N$ is very small.

  So the proportion gets very close to 1.

So almost all numbers …

## Presque tout

- There is an even more strange result: Choose any sequence of digits—say, 1066.

  Then 'almost all numbers contain that sequence'—like, for instance, 4001210660007. (*There is proof of this at the foot of the page**.)

- Although it is true that 'almost all numbers contain a 9', it is also the case (obviously?) that there is an infinity of numbers which do not contain a 9.

  But this infinity is swamped by the infinity of numbers with a 9.

- This is like the situation with Rational and Irrational numbers.

  Amongst the Real numbers there is an infinity of Rationals (terminating or recurring decimals) and an infinity of Irrationals (non-terminating and non-recurring decimals).

  The infinity of irrationals is 'infinitely greater' than the infinity of rationals (see pages 94-97).

  So 'almost all Real numbers are irrational'.

  To put this another way: suppose you have an infinite grid of infinitesimal dots (with unit spacing).

  Choose one of the dots.

  Almost all the lines through your chosen dot will never meet another dot!

*Write these numbers in base 10000 (the first power of 10 greater than 1066). In this base each number up to 9999 will be given a different symbol:
e.g. 10→Φ, 11→ľ, 12→Ч… 1066→ᑎ…
The proof about 9 could be adapted to prove that in base 10000 'almost all numbers contain ᑎ'.
Here is one of them: 4Чᑎ7.
Convert this number to base 10:
4Чᑎ7 → $10000^3 \times 4 + 10000^2 \times$ Ч $+ 10000 \times$ ᑎ $+ 7$
$= 4000000000000 + 1200000000 + 10660000 + 7$
$= 4001210660007$.
It contains the sequence 1066. So 'almost all numbers in base 10000 contain ᑎ' means that 'in base 10, almost all numbers contain 1066'.

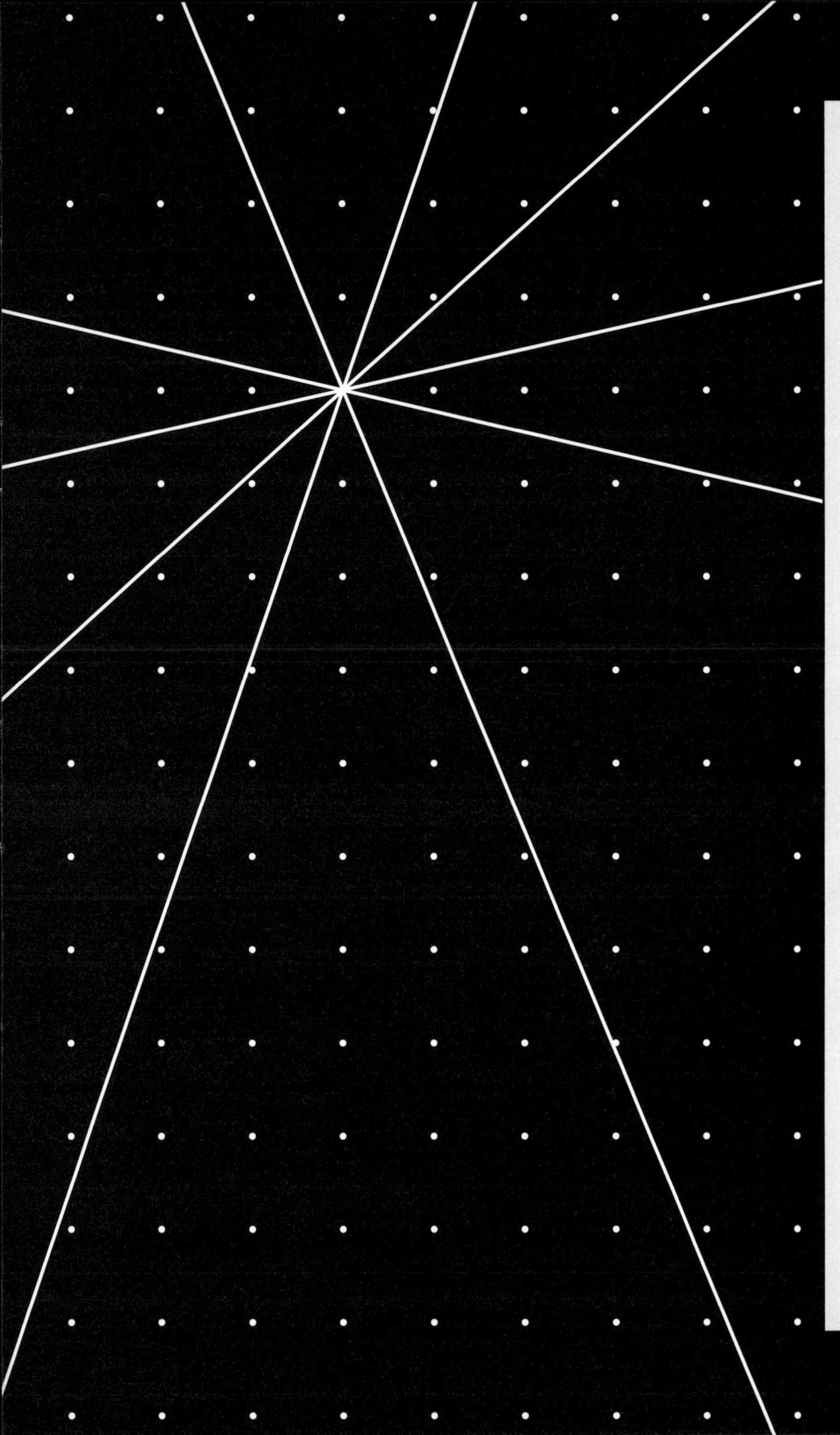

'It's a deal,' he said.

It amazed me that he did noi haggle. Only later was I to realise that he had entered my house with his mind made up to sell the book. Without counting the money, he put it away.

We talked about India, about Orkney, and about the Norwegian jarls who once ruled it. It was night when the man left. I have not seen him again, nor do I know his name.

I thought of keeping the Book of Sand in the space left on the shelf by the Wiclif, but in the end I decided to hide it behind the volumes of a broken set of The Thousand and One Nights. I went to bed and did not sleep. At three or four in the morning, I turned on the light. I got down the impossible book and leafed through its pages. On one of them I saw engraved a mask. The upper corner of the page carried a number, which I no longer recall, elevated to the ninth power.

I showed no one my treasure. To the luck of owning it was added the fear of having it stolen, and then the misgiving that it might not truly be infinite. These twin pro-occupations intensified my old misanthropy. I had only a few friends left; I now stopped seeing even them. A prisoner of the book, I almost never went out anymore. After studying its frayed spine and covers with a magnifying glass, I rejected the possibility of a contrivance of any sort. The small illustrations, I verified, came two thousand pages apart. I set about listing them alphabetically in a notebook, which I was not long in filling up. Never once was an illustration repeated. At night, in the meagre intervals my insomnia granted, I dreamed of the book.

Summer came and went, and I realized that the book was monstrous. What good did it do me to think that I, who looked upon the volume with my eyes, who held it in my hands, was any less monstrous? I felt that the book was a nightmarish object, an obscene thing that affronted and tainted reality itself.

I thought of fire, but I feared that the burning of an infinite book might likewise prove infinite and suffocate the planet with smoke. Somewhere I recalled reading that the best place to hide a leaf is in a forest. Before retirement, I worked in Mexico Street, at the Argentine National Library, which contains nine hundred thousand volumes. I knew that to the right of the entrance a curved staircase leads down into the basement, where books and maps and periodicals are kept. One day I went there and, slipping past a member of the staff and trying not to notice at what height or distance from the door, I lost the Book of Sand on one of the basement's musty shelves.

A quadrilateral is given a half-turn about midpoint 1.

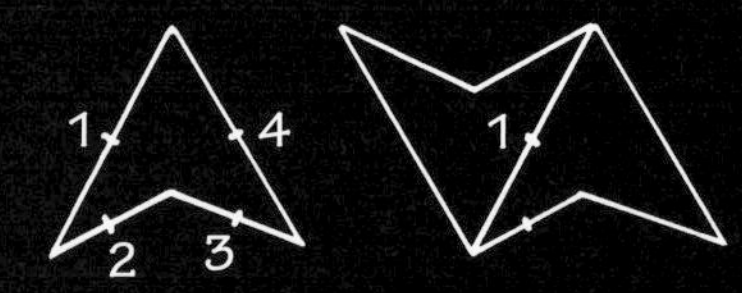


These two are half-turned about midpoint 2.

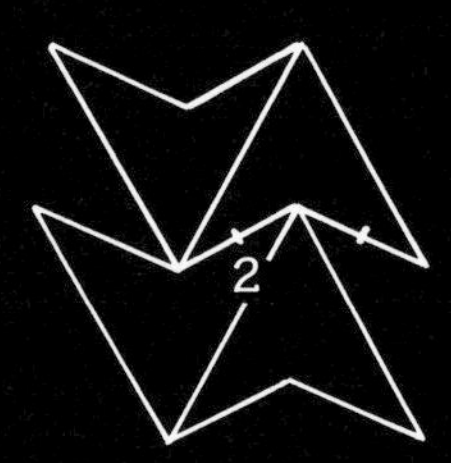


These four are half-turned about midpoint 3.

These eight are half-turned about midpoint 4.

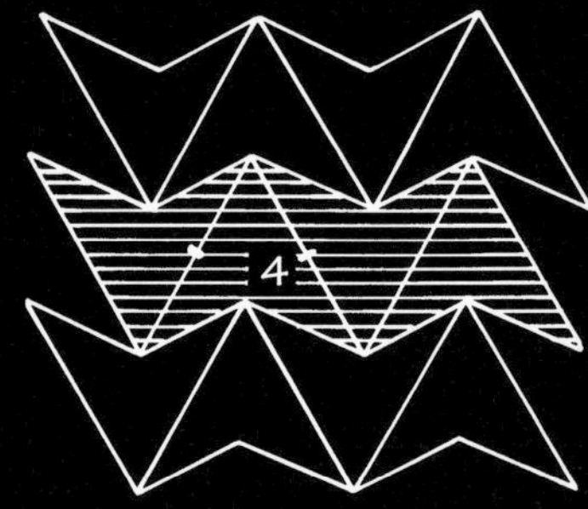


There is now an overlap.

But the tessellation still grows.

There are twelve quadrilaterals.

These are half-turned about midpoint 1, and so it goes on successively half-turning about midpoints 2, 3, 4, 1 …

In the drawing the initial position of the quadrilateral is at the light centre.

Although there is always some overlap, a new part of the plane is covered at each stage.

So the number of quadrilaterals grows:

1, 2, 4, 8, 12, 18, 24, 32, 40, 50, 60 …

Eventually …

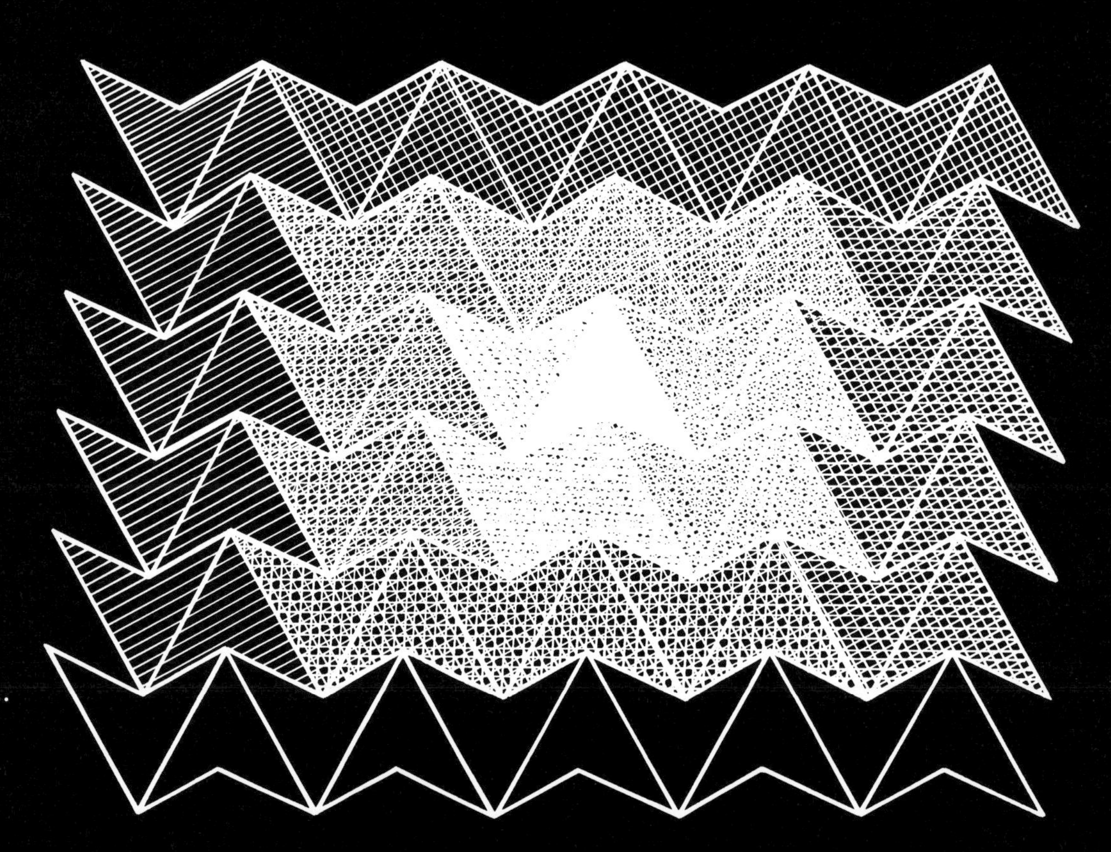

## How long is the coastline of the Isle of Wight?

The coastline has been measured off maps of various scales by 'stepping' round with steps of constant length (0.4 cm).

The conversions to actual lengths (in km) are shown in the table below.

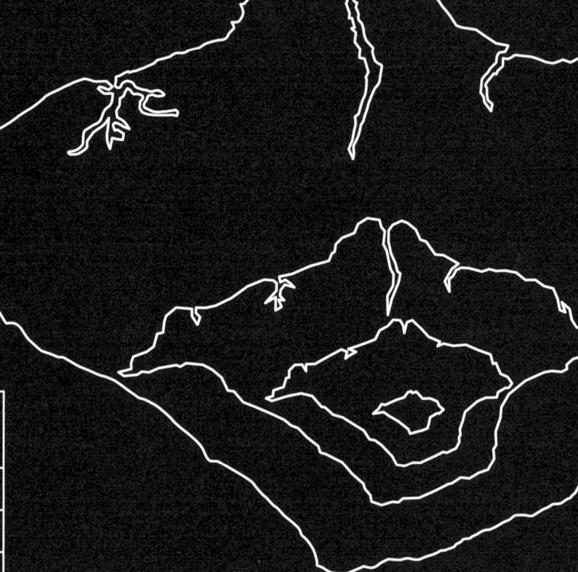

| Scale of map (1 cm = ) | Length of a 0.4 cm step | No. of steps | Length of coastline |
|---|---|---|---|
| 40 km | 16 km | 5 | 80 |
| 10 km | 4 km | 23 | 92 |
| 5 km | 2 km | 52.5 | 105 |
| 2.5 km | 1 km | 119 | 119 |

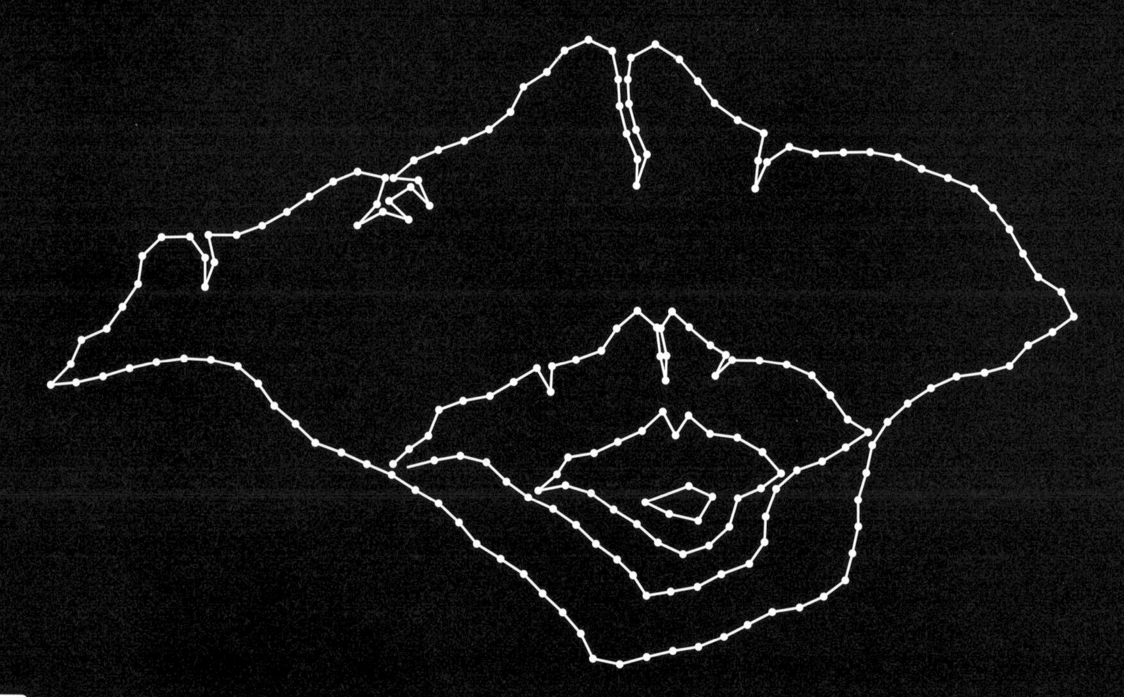

The smaller the steps, the closer we get to the actual coastline, so the measurement gets more accurate.

But the table suggests that as the length of the steps gets smaller the length of the coastline gets steadily greater.

Is there a limit to this—or is the coastline infinitely long?

The model opposite indicates that the latter is not impossible.

The four sides of a rhombus …

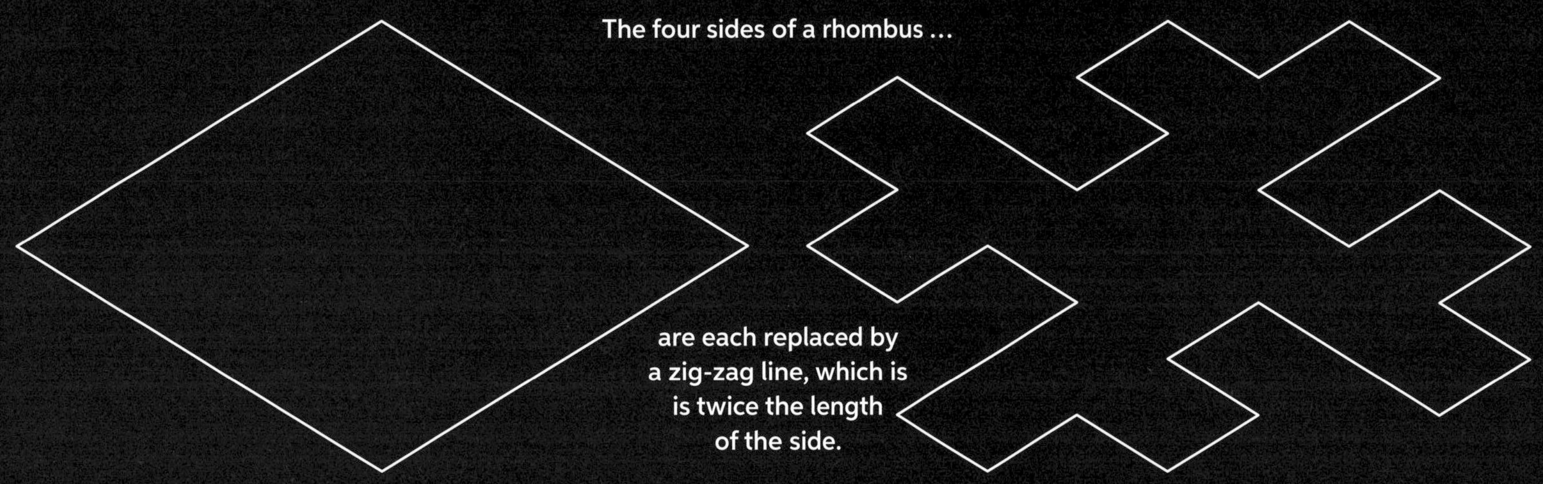

are each replaced by
a zig-zag line, which is
is twice the length
of the side.

So in Stage 1 the perimeter is doubled.

Each line-segment in Stage 1 is now replaced by a similar zig-zag line, thus doubling each line segment.The perimeter is again doubled so in Stage 2 it is now four times that of the rhombus.

The same kind of zig-zag replaces each of the stage 2 line-segments. The perimeter is now eight times that of the rhombus. At stage 4, it will be 16 times, and the process can go on indefinitely, doubling the perimeter each time.

Of course the ultimate form (which is called a fractal) can only be imagined.

Likewise the meaning of the statement: ‘The fractal has an infinitely long perimeter’ refers to an idea rather than a physical measurement.

## Cantor sets

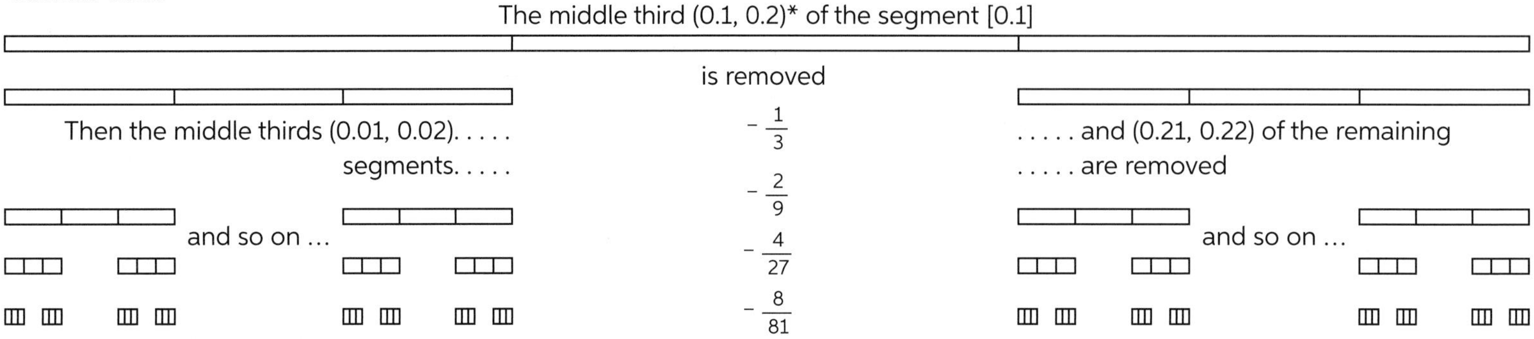

The middle third (0.1, 0.2)* of the segment [0.1]

is removed

Then the middle thirds (0.01, 0.02). . . . . segments. . . . .

. . . . . and (0.21, 0.22) of the remaining . . . . . are removed

$-\frac{1}{3}$

$-\frac{2}{9}$

$-\frac{4}{27}$

$-\frac{8}{81}$

and so on …

and so on …

What, ultimately, is left?

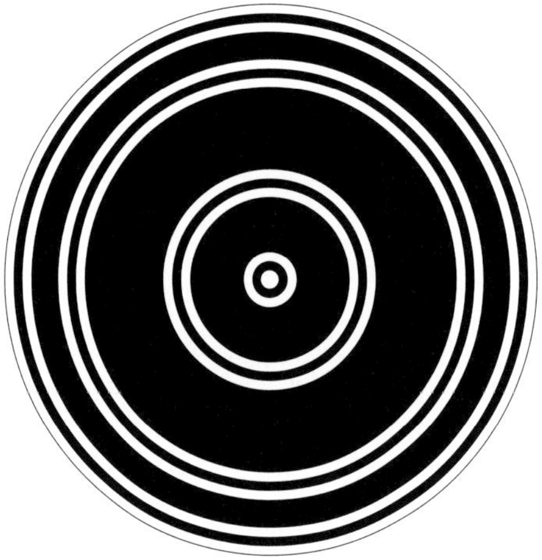

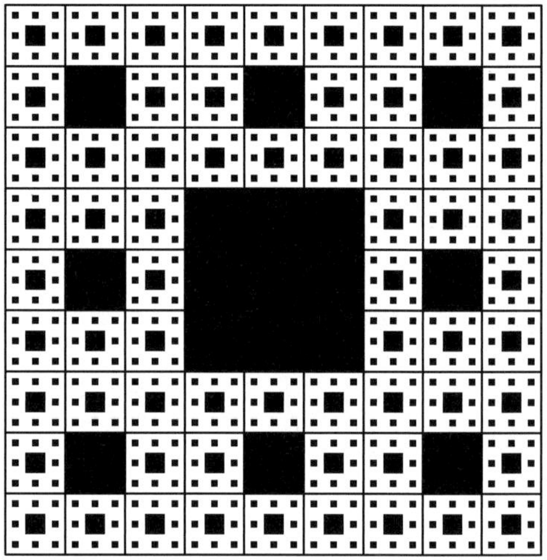

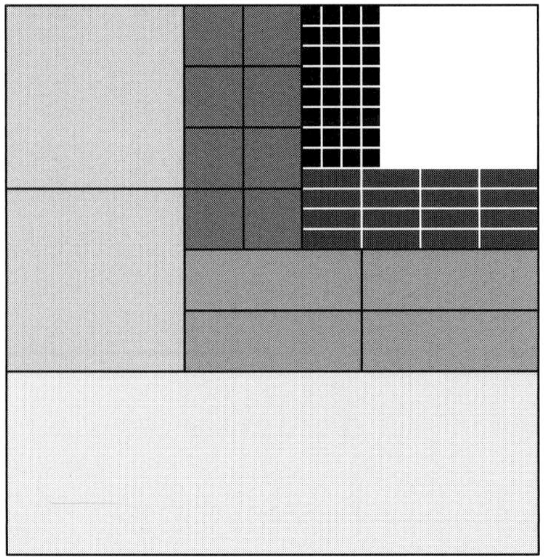

Here the 'middle ninth' of the square has been blacked out. And then the middle ninths of each of the eight remaining ninths. And so on.

Do you find this statement credible?

Ultimately the whole square would be black. All of it, that is, except infinitesimally thin white lines outlining each black square.

*( ) denotes an open set (not including the end points); [ ] means a closed set. Numerical notation here is in tricimals: 0.1 = 1/3, .01 = $1/3^2$ etc.

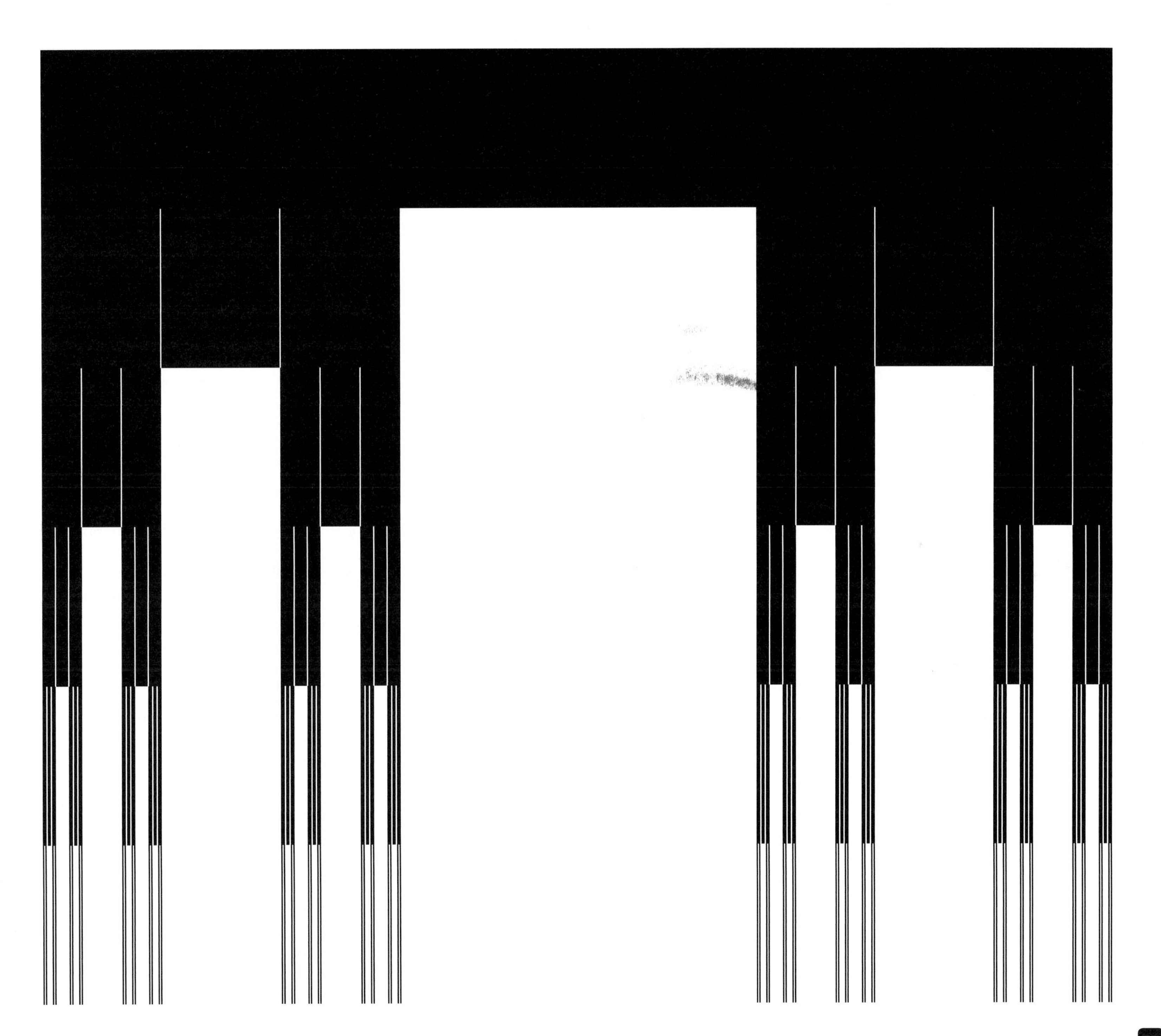

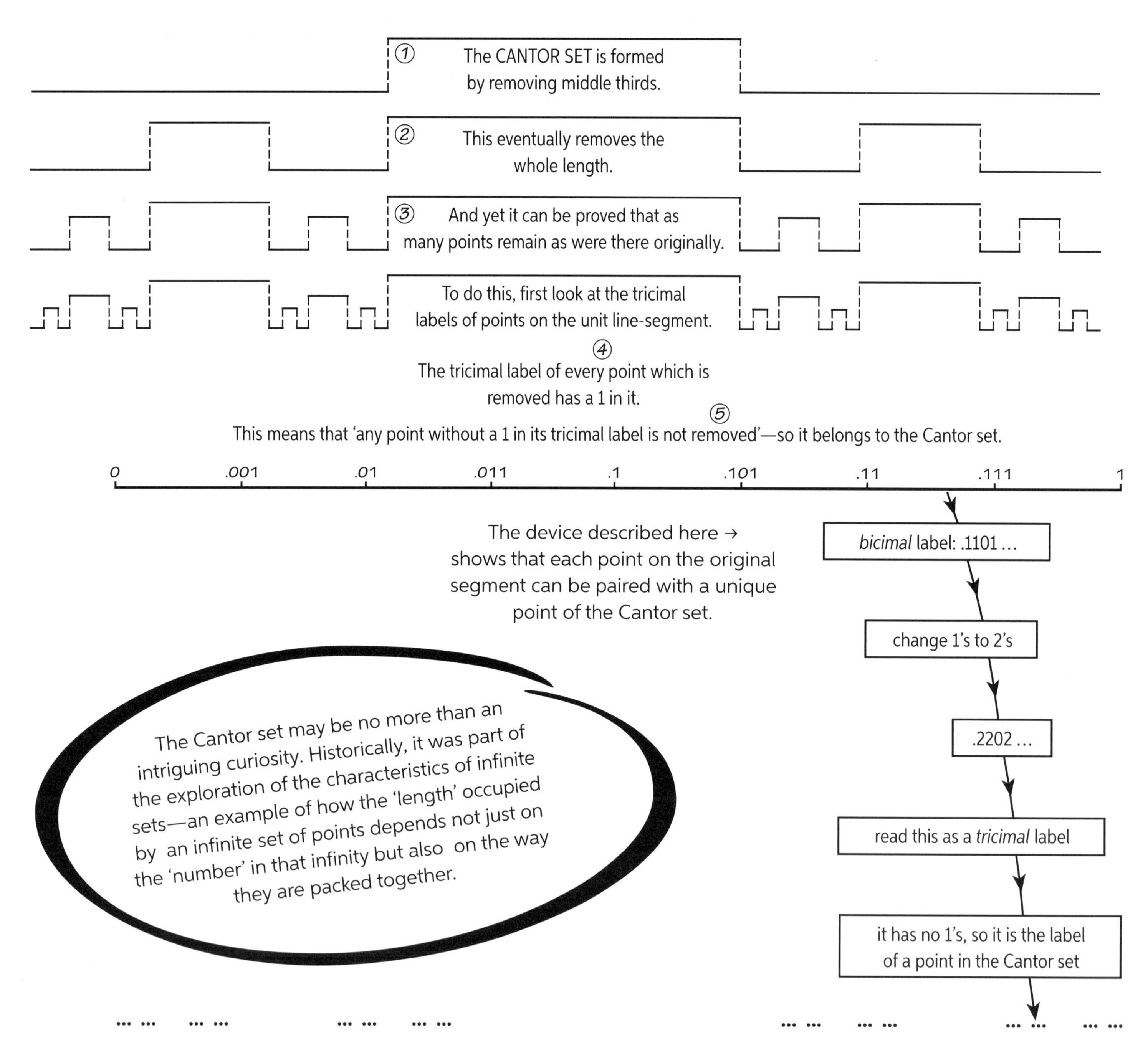

① The CANTOR SET is formed by removing middle thirds.
② This eventually removes the whole length.
③ And yet it can be proved that as many points remain as were there originally.
To do this, first look at the tricimal labels of points on the unit line-segment.
④ The tricimal label of every point which is removed has a 1 in it.
⑤ This means that 'any point without a 1 in its tricimal label is not removed'—so it belongs to the Cantor set.
0
.001
.01
.011
.1
.101
.11
.111
1
The device described here → shows that each point on the original segment can be paired with a unique point of the Cantor set.
bicimal label: .1101 …
change 1's to 2's
.2202 …
read this as a tricimal label
it has no 1's, so it is the label of a point in the Cantor set
The Cantor set may be no more than an intriguing curiosity. Historically, it was part of the exploration of the characteristics of infinite sets—an example of how the 'length' occupied by an infinite set of points depends not just on the 'number' in that infinity but also on the way they are packed together.
The whole line is removed—but as many points are left as were there to start with.

① Does it have to be thirds that are taken out? What happens if you remove a quarter each time? or a tenth? or a millienth? Does the argument still hold?

② How much is kept?

① → [× 2/3] → 2/3, 4/9, 8/27, 16/81 ··· getting smaller and smaller – until nothing is left.

③ What are the points that are left? – the boundary points. Only 'open sets' are removed – ie, all the points between 1/3 & 2/3, leaving just two end points.*

What does "as many" mean? For finite sets we can count them – ie, pair them off with the counting numbers. For infinite sets, it means we can pair off the members of one set with those of the other.

– That's what the rest of this argument is about – arranging this pairing.

④ Tricimals – like 'decimals' but working in base 3 ('Bicimals': in base 2)

⑤ The tricimal of the left hand boundary-point of every removed 'third' ends in a 1.

⑥ Logic is a bit tricky here. Compare it with this: If it were true that 'every person sent to prison (is removed) has committed a crime (has a 1 in it)' – then it would follow that any person who has not committed a crime is not sent to prison. (However, it would not follow that every person who has committed a crime is sent to prison.)

* At first sight it seems that every point in the Cantor set is describable by a fraction whose denominator is a power of 3. For example, 1/3, 2/3, 2/3 + 2/9, 2/9 + 2/81 + 2/729, etc.
But though this is true at all finite stages in the construction, it does not remain true at the infinite limit.
For instance; the recurring tricimal ·020202.. is a Cantor set point (it has no 1 in it). What is it's value? In fact, this condition – a tricimal without a 1 – means that all manner of unexpected numbers – rational and irrational – are included.

# HOTEL INFINITY

There was once a hotel in the mountains. So many people liked to go there that the manager decided to extend the hotel.

But still it was always full so he continued to add until eventually... it became infinitely large. The manager was very pleased and he especially liked his slogan...

'We are always full – but we always have room for you!'

One day a gentleman came to the hotel asking for a room. The receptionist looked through her books but could not find an empty room in the hotel. But the manager knew what to do. He gave orders for all the guests to be moved into different rooms.

The guest in room 1 was moved to room 2. The guest in room 2 was moved to room 3. The guest in room 3 was moved to room 4 and so on.

Everyone had a room to move to – because although there was an infinite number of guests, there was also an infinite number of rooms. And of course room 1 was then empty and that was given to the new arrival.

The other guests were not very pleased at having to change rooms.

And anyway they didn't much like the manager, who seemed a bit of a know-all. So they were very excited the next day when one of the buses from the 'Infinite Coach Hire Company' unloaded its passengers at the hotel. 'We always have room for you' one of them mimicked the manager, and the others laughed.

Seeing this infinite crowd of new guests, Miss Keeper at once called the manager. He thought for a moment and then he told her, 'Move the guest in room 1 into room 2 and the guest in room 2 into room 4 and the guest in room 3 into room 6 and so on. This will leave vacant rooms 1, 3, 5, 7, etc. Put the new arrivals into those rooms.'

The guests were very upset at having to move rooms again, but they were even more upset that the manager had again found a way of doing the impossible. So they called a meeting to think up an idea that would really beat him. No one could come up with anything that was foolproof until at last the professor had an idea which he thought would work:

**The professor's plan**

1. During the next few weeks there will be a series of meetings.
2. First there will be an"empty meeting" with no one present.
3. Next, every guest will go to a meeting at which only he, or she, is present.
4. Then the guests will meet in pairs: these meetings must cover all possible pairs of guests.
5. Then there will be meetings of all possible groups of three guests.
6. Then meetings of four guests, then of fives, then of sixes, and so on, until-
7. There have been meetings of all possible groups of all sizes.
8. This series of meetings must be completed by 1st. december.
9. The purpose of each meeting is to invite a guest (who may, or may not be a present guest) to stay at this hotel next Christmas.
10. When a meeting has agreed on its guest, the manager must be asked to reserve a room for that person.
11. No two meetings may invite the same person (a list will be kept to avoid this happening)
12. The professor will decide whom the"empty meeting"(rule 2) will invite, and he will ask the manager to book a room for that guest.

Although the other guests didn't really understand it, they thought it must be a very good plan. The rules were sent to all the guests.

The meetings quickly got under way. Some of the guests wanted to stay on for Christmas and they were 'invited' by one of the meetings. But when he booked the rooms, the manager did not always give these guests the rooms they already occupied. And then one day a member of staff brought him a copy of the professor's plan which she had found whilst cleaning one of the rooms. The plan worried him. That night he couldn't sleep. He went to his office and began to write.

I'll label each guest with the number 'of the room they now occupy. So Mr. Zawirski – he's in room 26 – he'll be just '26'. The person in room N will be called 'N'... now these meetings – I'll label them according to the number of the room I give to its guest. The meeting whose guest has Mr. Zawirski's room will be $M_{26}$'.

The meeting whose guest will occupy room N will be called $M_N$. Then he thought about the time when he would have made all the room allocations. 'Some of the people will have been to the meeting whose guest I am putting into the room they now occupy. But I had to give a room (it was 126) to the guest of the empty meeting and person 126 was not at that meeting – no one was.

So there's at least one person who does not go to the meeting; whose guest gets that person's room.

There may be more than one like that. And they will have a meeting (rule 7) – and I shall give that meeting's guest a room – say 314. So it's meeting $M_{314}$. Now what about person 314?

Was she (or he) at meeting $M_{314}$? No – because $M_{314}$'s guest doesn't get the room of anyone at $M_{314}$.

But $M_{314}$ is a meeting of all the people who did not go to the meeting whose guest is given their own room. That means 314 would have gone to $M_{314}$.

'But that's not possible' the manager cried out. It is saying, 314 is *not* at the meeting means that 314 *is* at the meeting. He (or she) can't be both *at* the meeting and *not at* the meeting.

So I can't give this meeting's guest room 314 or any other room!' The manager thought all night about this – but he could find no way out. So in the morning he wrote a letter of resignation and left the hotel for good.

The guests watched him go, and then had a little celebration at their victory.

But after a few weeks they found the service at the hotel began to get worse and worse. They had to wait ages for their meals, and sometimes the potatoes ran out altogether.

And there were awfull muddles about the bookings for rooms. So they began to feel very sorry that they had been jealous just because the manager was so clever.

*The remaining pages of this book are devoted to an outline of the ideas which culminated in Cantor's theory of transfinite numbers.*

Galileo pointed to the following paradox:

**Every number has a square**

1 2 3 4 5
↓ ↓ ↓ ↓ ↓
1 4 9 16 25

So there are as many squares as numbers.

**BUT**

**?**

**Many numbers are not squares**

2 6 10 11 14 13 15 4 16 9 25 18 21 3 8 7 12 17 19

So there are more numbers than squares.

Bolzano pointed out that this kind of paradox is common. Here are two of his examples:

**1**

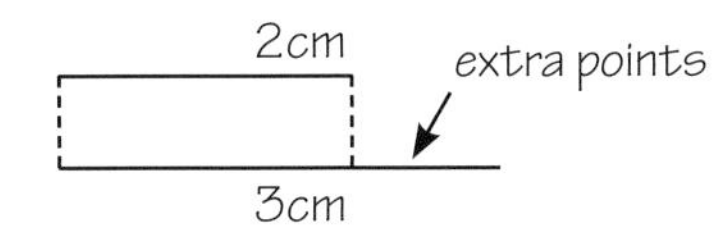


There are more points in a 2cm line than in a 3cm line.

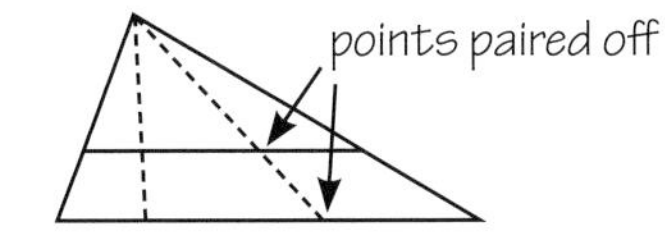


BUT The points can be paired off so there is the same number of points in the two lines.

**2** Numbers between 0 & 5

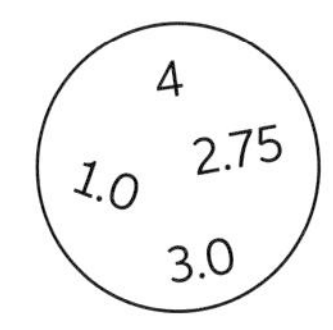


form a *part* of numbers between 0 & 12. BUT

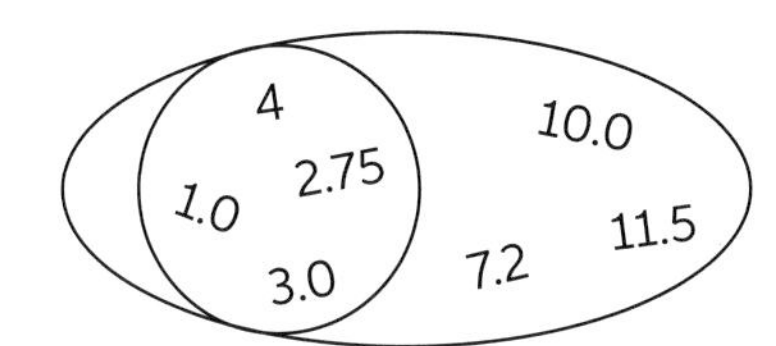


So there are *more numbers* between 0 & 12 than between 0 & 5.

Any number in the '5-set' can be made into a number in the '12-set.'

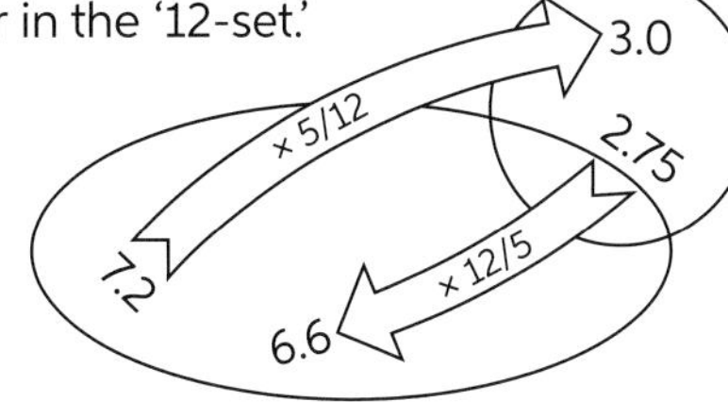


and any number in the '12-set' can be made into a number in the '5-set.'

So because the numbers can be paired off like this there are the *same number* of numbers in the two sets.

Whenever you are trying to decide which is the bigger of two infinite sets you are liable to run into the kind of paradox that Galileo and Bolzano pointed out.

Here are three ways out:

You could say:

**With infinite sets you can't compare their sizes.**

Or you could say:

**Infinite sets are not like finite sets—in an infinite set you can have a part of it which is 'equal' to the whole of it.**

Or you could say:

**Counting by 'pairing-off' does not work for infinite sets.**

Bolzano chose the last of these escape-routes.

He argued that, for finite sets, pairing-off tells you the numbers in two sets are equal if, when you come to the end, there are none left over in either set. But for infinite sets you can never get to the end, so it does not tell you anything.

So he said there are different-sized infinities: The infinity of numbers between 0 and 12 is greater than the infinity of numbers between 0 and 5. The infinity of points on a 2cm line is less than the infinity of points on a 3cm line.

Bolzano (1781–1848)

This seems very sensible. This extract from an 18th-century encyclopaedia, suggests that it was the accepted point of view even before Bolzano's time.

INF

If an *infinitely* ſmall quantity be multiplied into an *infinitely* great one, the product will be a finite quantity.
INFINITELY *Infinite Fractions*, or all the powers of all the fractions whoſe numerator is one, are together equal to an unit. See the demonſtration hereof given by Dr. Wood, in Hooke, *Phil. Coll.* N° 3. p. 45, *ſeq.*
Hence, it is deduced, 1°, That there are not only *infinite* progreſſions, or progreſſions *in infinitum*; but alſo *infinitely* farther than one kind of infinity. 2°, That the *infinitely* infinite progreſſions are notwithſtanding computable, and to be brought into one ſum; and that not only finite, but ſo ſmall as to be leſs than any aſſignable number. 3°, That of *infinite* quantities, ſome are equal, others unequal. 4°, That one *infinite* quantity may be equal to two, three, or more other quantities, whether *infinite* or finite.
INFINITE *Series*. See the article SERIES.
INFINITIVE, in grammar, the name of one of the moods, which ſerve for the conjugating of verbs

Cyclopaedia by E.Chambers, published 1751.

But later mathematicians preferred the second escape-route, even though it seems less plausible—that there are as many numbers between 0 & 1, for instance, as between 0 & 100.

And it was just that kind of thing which they used to *define* an infinite set.

A German mathematician, Dedekind, first took this bull by the horns.

In 1872, he wrote:

A system S is said to be infinite when it is similar to a proper part of itself; in the contrary case S is said to be a finite system.

'system' = set.
'a proper part' = a subset which does not contain all the elements of the set.
'similar' = elements can be paired off.

Dedekind's notion of infinite sets being similar (when you can pair off their elements) raised many questions.

Dedekind (1831–1916)

For example:

Can we say similar sets are equal in number?

If so, what kind of numbers are these 'infinities'?

Are some infinite numbers larger than others?

Or are all infinite sets similar to each other?

Can infinite numbers be added together?

If so, does this produce larger infinities?

...

and so on

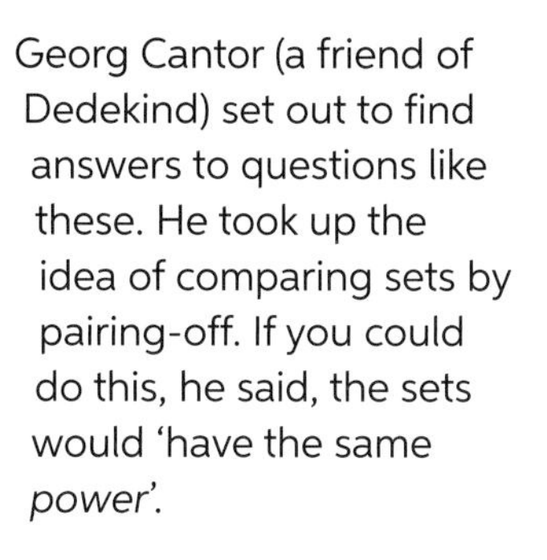

*Cantor (1845–1918)*

Georg Cantor (a friend of Dedekind) set out to find answers to questions like these. He took up the idea of comparing sets by pairing-off. If you could do this, he said, the sets would 'have the same *power*'.

But since you can't actually do this 'to infinity' you need a rule to tell you how you *could* do it.

What sort of rule?

ATTENSHUN!
FORM LINE!
FROM THE RIGHT,
NUMBER!

For a finite set you need to put the elements in some sort of order so that you can pair them off with the counting numbers.

The number of the last one is then the number of elements in the set.

For infinite sets there is not 'a last one'. But you can have a rule which tells you how to do the pairing off—for ever.

And Cantor saw that this rule would depend on being able to put the elements into a definite *order*.

If you can put the elements of an infinite set into some definite order then you can pair them off with the counting numbers.

Cantor called such sets *countable.*

| | 1 | 2 | 3 | 4 | 5 | 6 | 7 | 8 | 9 | 10 |
|---|---|---|---|---|---|---|---|---|---|---|
| odd numbers | 1 | 3 | 5 | 7 | 9 | 11 | 13 | 15 | 17 | 19 |
| square numbers | 1 | 4 | 9 | 16 | 25 | 36 | 49 | 64 | 81 | 100 |
| multiples of 10 | 10 | 20 | 30 | 40 | 50 | 60 | 70 | 80 | 90 | 100 |
| integers | 0 | -1 | +1 | -2 | +2 | -3 | +3 | -4 | +4 | -5 |
| fractions | | | | | | | | | | |

WHO'LL GO FIRST?

ME!

ME!

ME!

$\frac{1}{2}$ $\frac{3}{8}$ $\frac{1}{3}$ $\frac{1}{1000000}$ $\frac{11}{500}$ $\frac{7}{5}$ $\frac{5}{3}$ $\frac{1}{1000}$ $\frac{1}{1000000000}$

D'you reckon this lot is countable?

Doesn't look like it. They can't get themselves in order – don't even know where to start.

INFINITE SETS which can be put into 1-to-1 correspondence with the counting numbers are said to be COUNTABLE

More squares or more circles?
About the same, I'd think. Let's see, I'll pair them off.
Oh – there's one more circle.
Now here's one for you – more black blobs or white blobs?
More black blobs obviously.
Sure? – count them.
There – lots left over.
But you haven't finished – the white blobs go on for ever. So do the black blobs.
One million, five hundred and fifty-seven thousand two hundred and twelve – one million, five hundred and fifty-seven thousand two hundred and thirteen – one mill…

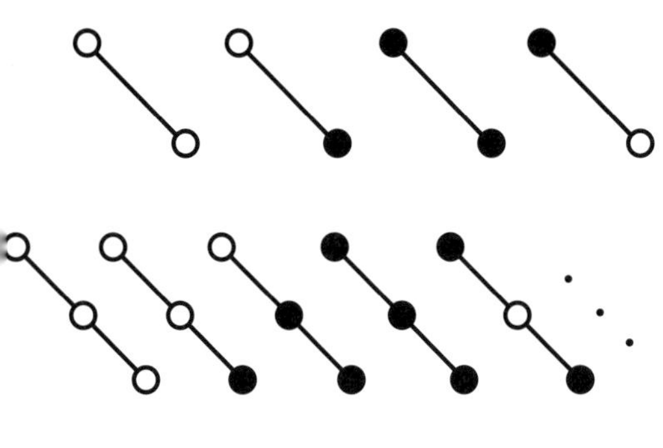

**That gives me an idea. You know those fractions which we couldn't get in order. Well, we could use these blobs.**

For fractions?

**Yes. Give them coordinates. Like (2,2) or (5,4). Then you could write them as though they were fractions – 2/2 and 4/5.**

So?

**Well then each blob has a name – a fraction. And each fraction has a blob.**

All right.

**Now we can put the blobs in order with that np-and-down system. So we can put the fractions in order – 1/1, 2/2, 1/2, 1/3, 2/3, 3/3, 4/4, 3/4.**

Great! And you won't miss out any because each fraction has a blob.

**Right. And if we've got them in order we can pair them off with the counting numbers – the white blobs. So, the fractions are countable.**

You're a genius!

**Well, it's just a little idea...**

Oh no you're not a genius. I've thought of a snag. In your system there are no fractions greater than 1. Where's 5/4, for instance?

**Oh yes. That's up here. above the sloping line. All the fractions greater than 1 are up there. We shall have to put black blobs all over.**

And then you can't number them the same way because once you start going up a column, you'll never come down again.

**Oh dear, yes. That needs thinking about.**

*Well, you've got a good system to make sure you don't miss out any black blobs – but you see, you can always find another white blob to pair off with each black one – If you found there wasn't one, that would mean you'd come to the end of the line of white blobs. But it goes on for ever!*

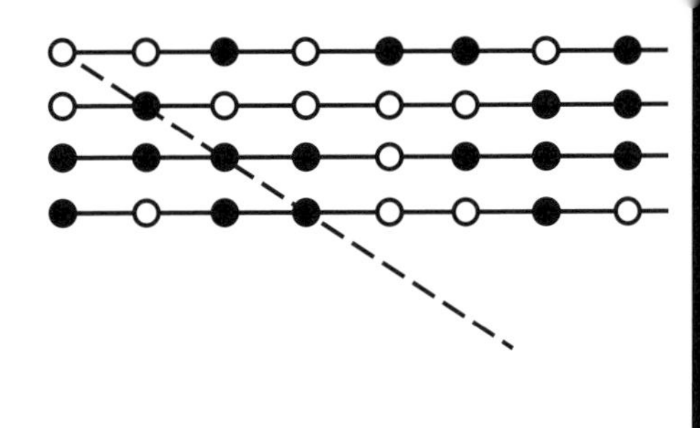

Cantor showed that it is possible to put *all* the fractions in an order so that you can 'count' them, leaving none out. Here is his system:

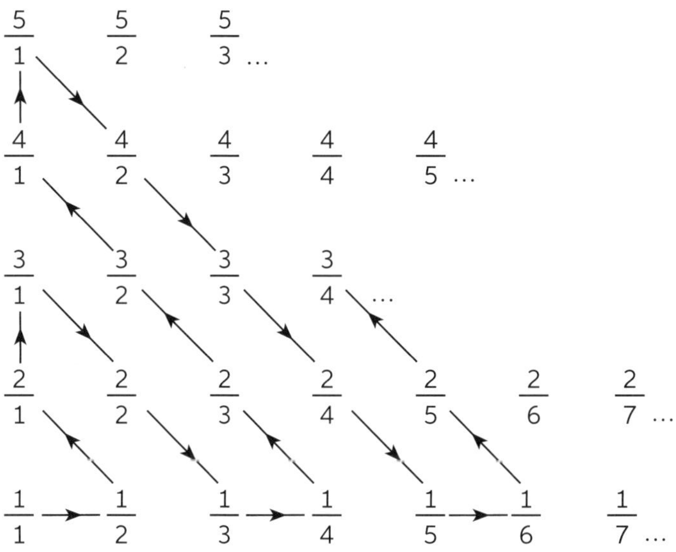

$\frac{5}{1}$ $\frac{5}{2}$ $\frac{5}{3}$ …

$\frac{4}{1}$ $\frac{4}{2}$ $\frac{4}{3}$ $\frac{4}{4}$ $\frac{4}{5}$ …

$\frac{3}{1}$ $\frac{3}{2}$ $\frac{3}{3}$ $\frac{3}{4}$ …

$\frac{2}{1}$ $\frac{2}{2}$ $\frac{2}{3}$ $\frac{2}{4}$ $\frac{2}{5}$ $\frac{2}{6}$ $\frac{2}{7}$ …

$\frac{1}{1}$ → $\frac{1}{2}$ $\frac{1}{3}$ → $\frac{1}{4}$ $\frac{1}{5}$ → $\frac{1}{6}$ $\frac{1}{7}$ …

[Can you be sure that all possible fractions ($\frac{37}{135}$, $\frac{5319}{64}$ … for instance) would be included?]

So he had shown that the set of fractions—the positive rationals, as they are called—is countable.

It has the same power as the set of counting numbers.

Just how surprising this is appears when you think 'how many' rationals there are—just between 0 and 1, say:

If the first dot is labelled 0 — and the last one is labelled 1

0 1

these two dots are 63/100 & 64/100

Put them under a magnifying glass and you will find another 99 dots between them!

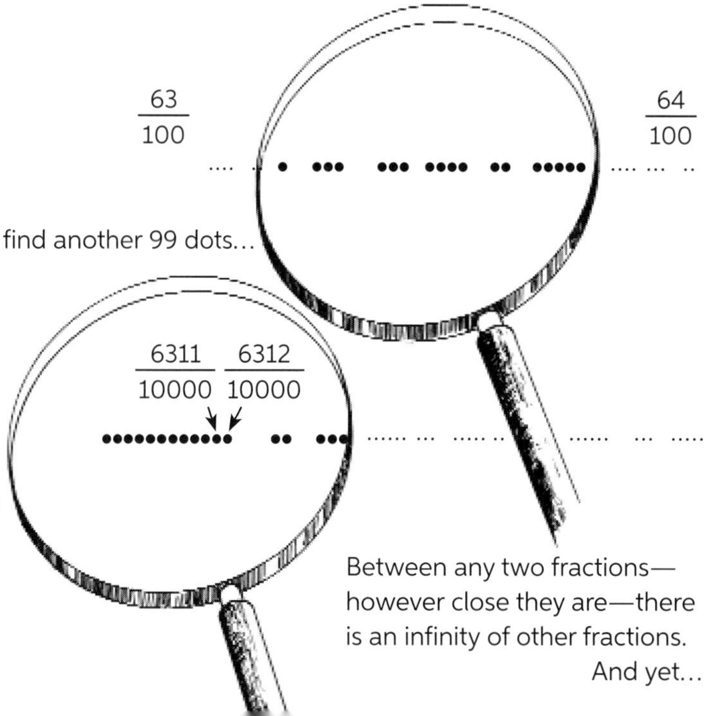

$\frac{63}{100}$ $\frac{64}{100}$

and between any two dots of these dots you can find another 99 dots…

$\frac{6311}{10000}$ $\frac{6312}{10000}$

and so on

and so on

Between any two fractions—however close they are—there is an infinity of other fractions.

And yet…

## A counting problem

Perhaps all infinities are the same …?

No — here is one that cannot be ordered and so it is not the same as the infinity of counting numbers.

We are making strings of black and white beads.

How many different strings can we make with just two beads?

with three beads?

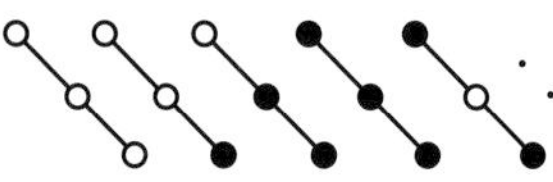

Now think about unending strings!

Can we put them in some order (so that we can 'count' them)?

Well, suppose we have done this:

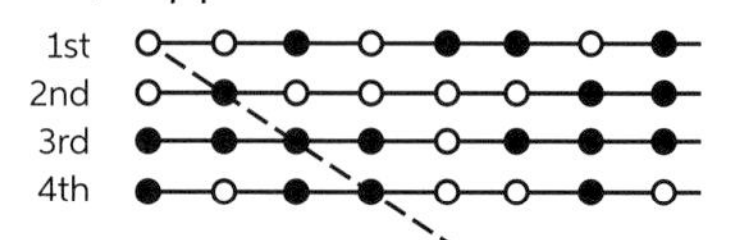


Can they all be here?

No—there is at least one missing!

Here is how to make it:

Pick out the diagonal line of beads

and in this change every black one into a white one and every white one into a black one.

This new string

is not the same as the 1st one because its 1st bead is different;

it is not the same as the 2nd one because its 2nd bead is different;

it is not the same as the 3rd one because its 3rd bead is different …

and so on …

It is different from every string in our list.

So our list was not complete.

(Perhaps we could put it in somewhere—at the beginning, say.

But then we could make another new string in the same way.)

So we can never make a complete list.

Infinite strings of beads don't seem very important.

But suppose you replace the black beads with the symbol 1 and the white beads with 0.

Imagine a line.

Imagine taking half of it by choosing to keep either the left- or the right-hand half.

Imagine taking half of this half, making a similar choice.

And so on …

The points at the 'beginning' of each selected half can be numbered 0 if it is a left half and 1 if it is a right half.

Some points like $1/4$, $3/8$, $13/16$

… have exact labels. But others always fall inside a 'half'—$1/3$ for instance:

This needs an infinite label—though it is often one with a repeating pattern.

Any point of the segment (0,1) which corresponds to an 'exact fraction' will have a label with a repeating pattern. These labels are called *bicimals*.

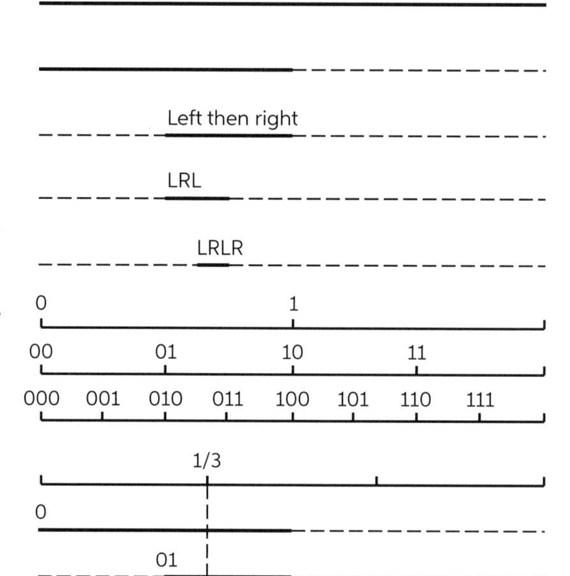


At one time it was thought—understandably—that any length would be expressible as an 'exact fraction' of any other length. But the Pythagoreans discovered that this was not true of the side and diagonal of a square.

This discovery led eventually to the concept of irrational numbers.

Unlike the infinity of fractions the infinity of infinite strings is uncountable.

The number $\sqrt{2} - 1$ is between 0 and 1.

Its bicimal form starts .011010101001 … and continues without ever settling into a repeating pattern.

Such numbers are called *irrational*.

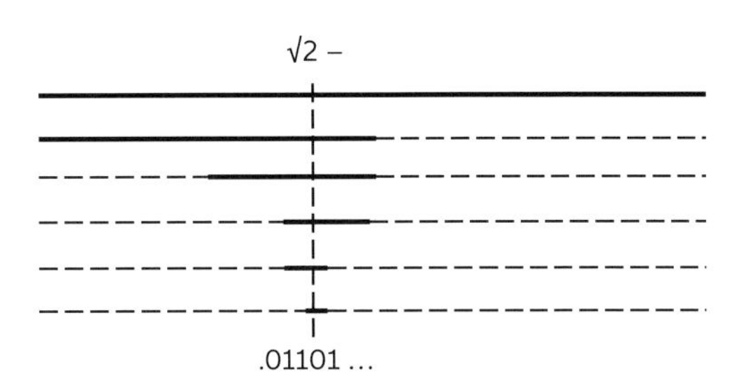


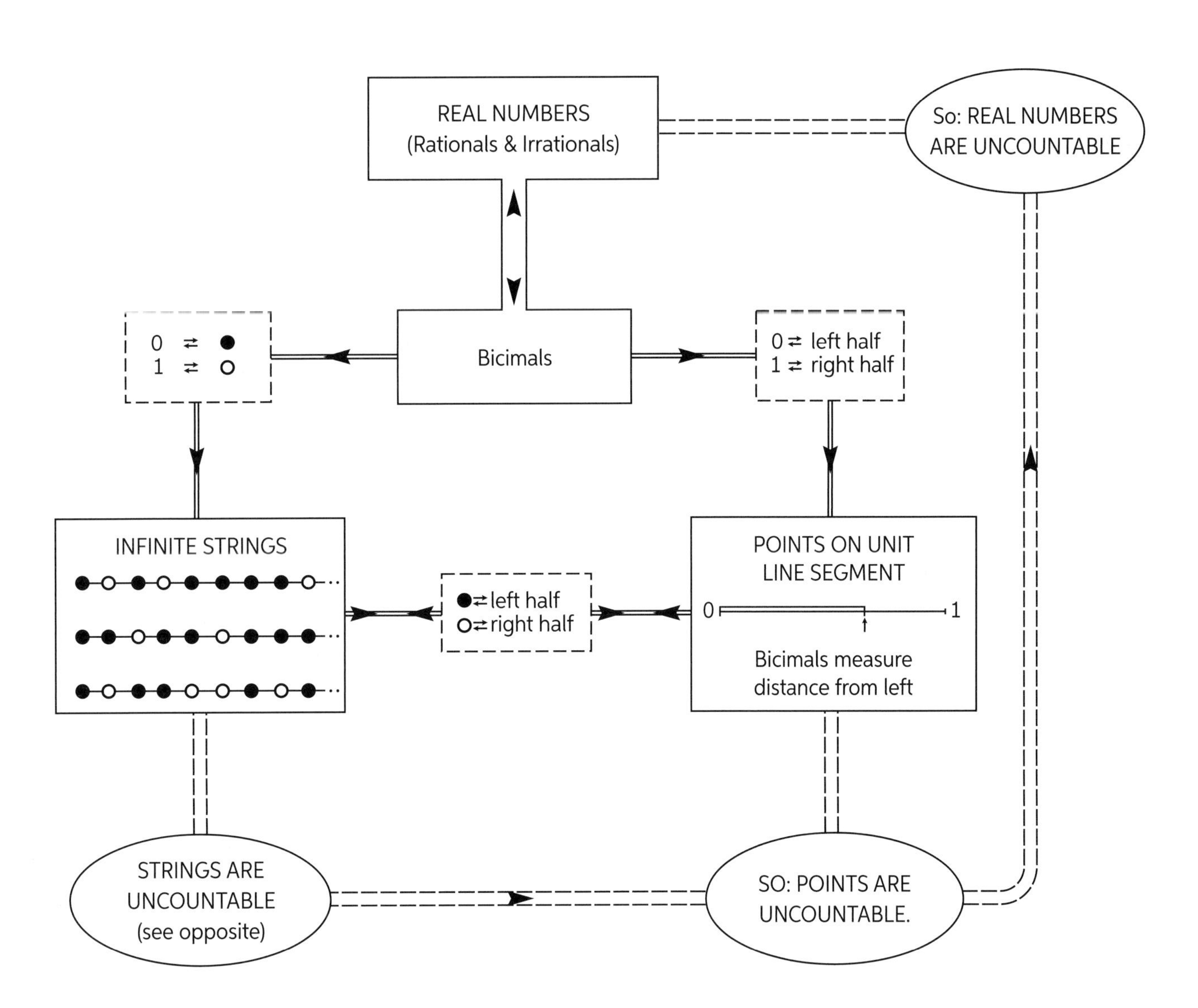

- Having shown that the rationals are countable Cantor tried to do the same for the set of all real numbers (irrationals included). The problem was to find a way of listing them so that they could then be matched with the counting numbers. He had to be sure, of course, that such a list would include them all.
- He could find no way of doing this; so he set out to prove that it was impossible to produce such a list. He devised the ingenious 'diagonal method' which we used on the string of beads.
- First he supposed that such a list had been made →

  .0 1 4 2 7 9 3 5 …

  .2 7 8 5 1 3 9 6 …

  .1 3 7 0 0 0 0 0 …

  .6 2 8 9 1 4 7 2 …

  From this he took the sequence of digits from the diagonal: 0 7 7 9 …

  Then he made a new sequence by imagining every digit in the diagonal sequence to be changed.

  It might be 1 8 8 0 …

  With a decimal point in front, this would be a real number.

  Is it in the 'complete list'?

  It is not the 1st number because its 1st digit is different. It is not the 2nd number because its 2nd digit is different. It is not the 3rd number because its 3rd digit is different.

  And so on. It differs from every number in the list by at least one place.

  So it is a new number.

  So the list is not complete after all.

  And this applies to any list of reals. It is not possible to make a complete list.

  And without that you cannot pair off the real numbers with the counting numbers.

  **The reals are uncountable.**
- So a new infinity was discovered—the infinity of the continuum, it was called. This discovery opened up new questions:

  For example,

  are there bigger infinities still?

  are there infinities between that of the counting numbers and that of the reals?

  are these infinities 'numbers'?

  if so, how do they combine—can you add them, multiply them … ?

  To many of these questions Cantor found answers.

  But we leave them as questions.

If you have enjoyed this book, then there will be other Tarquin books and posters which would interest you. For our full range see www.tarquingroup.com. Our books and puzzles are available from bookshops, toyshops, art/craft shops or in case of difficulty directly by post.

For an up-to-date catalogue please write to

Tarquin, Suite 74, 17 Holywell Hill,
St Albans, AL1 1DT, UK

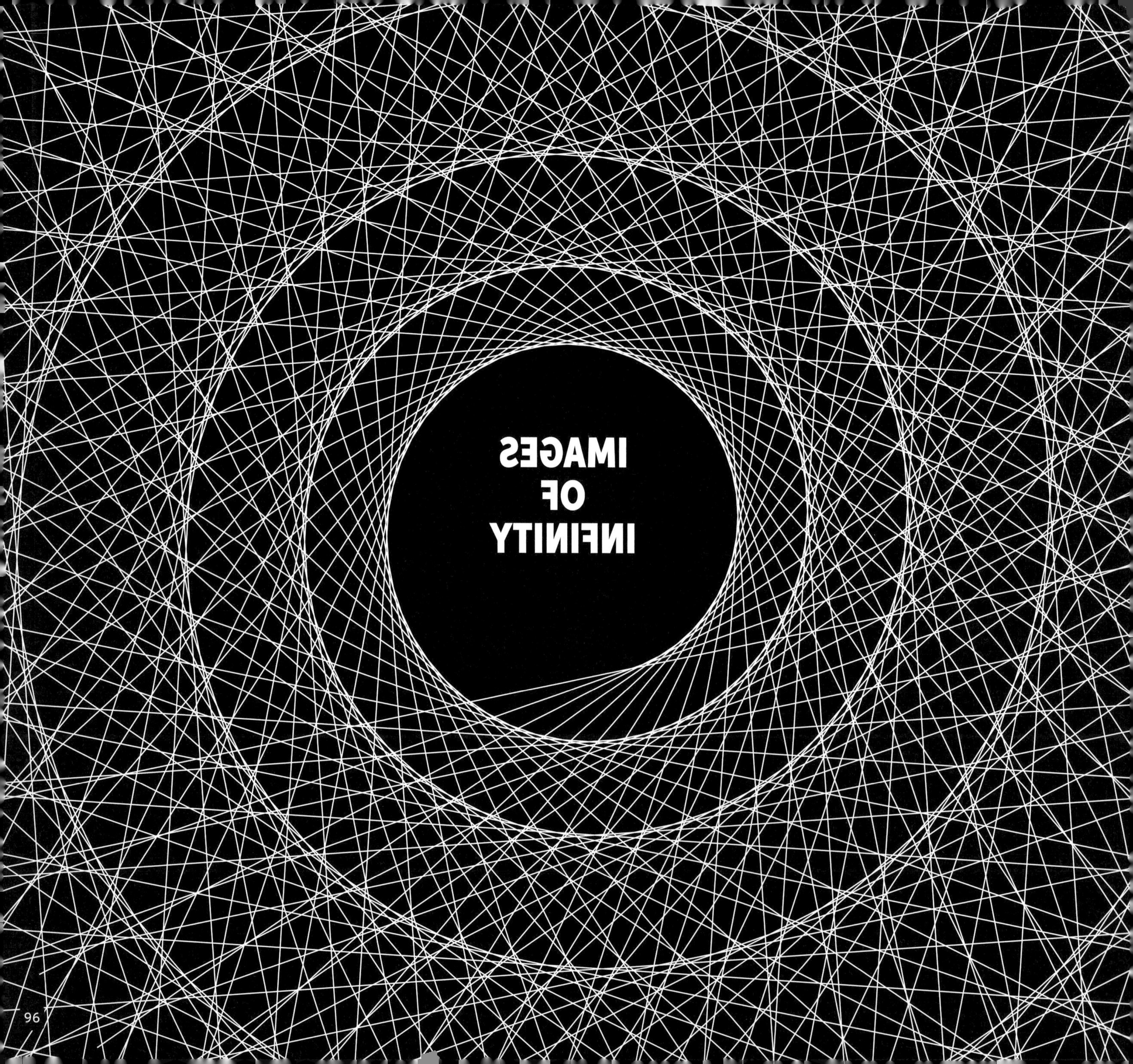
IMAGES
OF
INFINITY